Eduardo Pagel Floriano

Environmental, social and economic aspects of eucalyptus cultivation

Eduardo Pagel Floriano

Environmental, social and economic aspects of eucalyptus cultivation

subsidies for achieving sustainability

ScienciaScripts

Imprint

Cover image: www.ingimage.com

This book is a translation from the original published under ISBN 978-620-6-76039-9.

Publisher:
Sciencia Scripts
is a trademark of
Dodo Books Indian Ocean Ltd. and OmniScriptum S.R.L publishing group

120 High Road, East Finchley, London, N2 9ED, United Kingdom
Str. Armeneasca 28/1, office 1, Chisinau MD-2012, Republic of Moldova, Europe
Managing Directors: Ieva Konstantinova, Victoria Ursu
info@omniscriptum.com

Printed at: see last page
ISBN: 978-620-8-57093-4

Contents

PRESENTATION

Walking through the Pampa, fishing in sangas and trekking through forests were childhood activities that turned into a great passion for the natural environment, which was decisive in my choice of profession as a Forest Engineer; more recently they have led to my concern about man's relationship with the environment. Knowing that it is impossible to repeat with our children what we experienced with our parents is a hard realisation.

So it was in my early days on the Forestry Engineering course at UFSM, which I joined in 1975, that during one of Dr Franklin Galvão's lectures I learnt about the lack of prior environmental studies for the implementation of large projects. At the time I was just a young student and I couldn't understand how they hadn't yet created means of controlling the environmental impact of projects such as hydroelectric power stations, industries or even rural areas.

In 1986, after a few years of professional experience and due to my knowledge of statistics and computer programming, I was invited to take part in the analysis of the hundred or so forestry experiments that Cia Vale do Rio Doce had implemented in the North and Southeast regions since the 1960s, a programme developed in partnership with UFRRJ. During these activities, I learnt about the Bahia Sul project and had access to various documents, including technical feasibility studies. I even took part in a meeting, during the period of negotiations between Suzano and Vale, at Vale's headquarters in Rio de Janeiro, to provide information on the results of experiments with eucalyptus in the region covered by the project and on another occasion in Vitória, ES. The project for the new mill took advantage of the plantations carried out by the former Flonibra, which had been set up to build two mills (one in the north of Espírito Santo and the other in the south of Bahia), which after political negotiations resulted in the establishment of Cenibra in Minas

Gerais and the abandonment of the initial project. This involvement sparked my interest and the opportunity to work at Bahia Sul arose in 1989, when the RIMA for setting up the industry was approved and hires were made to complement the team of Forest Engineers. This was the start of the company's activities after the RIMA was approved. I was entrusted with the operational management of 1/4 of the forest areas and with the responsibility of setting up the infrastructure for that section, although half of the forests had already been planted by the forerunner Flonibra, I was able to witness the significance of an investment of 1.5 billion dollars, including forests and industry, in a region as impoverished as the south of Bahia. This was the volume of funds spent on the project until the end of the mill's adjustments in 1992, when it went into full operation.

The town of Mucuri, the seat of the municipality where the industry was set up, was just a fishing village by the sea with just one paved street, no basic sanitation, no hospitals, no secondary schools, and access to the BR-101 via a poorly maintained 40-kilometre dirt road. During the establishment of the factory, access to the town was asphalted, a landing field for medium-sized jets was built next to the factory, a residential village, an excellent hospital and an excellent 1st and 2nd grade school, and a telephone exchange was installed; today there is a well-equipped health centre in the town, the streets have been paved, sanitation is improving day by day and the other schools in the municipality have been improved and equipped.

But there is another side to this story. Nearby, between the Bahia Sul and Aracruz factories, there are many municipalities suffering negative socio-economic impacts. One of the most alarming cases is the municipality of Conceição da Barra, well known for having one of the most popular beaches in the north of Espírito Santo. The municipality lives mainly from eucalyptus cultivation, which takes up around 70 per cent of its cultivable land ; from sugar cane, whose alcohol and sugar industries own another significant chunk of land; from tourism; and from oil royalties exploited by Petrobras. But the population is poor. In the countryside it is minimal and employed either in sugar cane farming or forestry. Sanitation in the town is almost non-existent, the schools are poorly equipped, etc., etc. In this municipality, forestry is characterised as one of the main factors preventing sustainable development.

Pulp companies try to distribute taxes as much as possible between the municipalities where they operate, making the appropriate allocations and maintaining branches for tax purposes in all of them, even demanding that the service providers they hire also open branches and issue invoices where the services are rendered, but the biggest movement is due to turnover in the municipality where the industry is headquartered.

As a result, the advantages of pulp production are felt almost only in the municipalities where the industries are located and, although they generate wealth and benefits, they also generate poverty and dissatisfaction.

It's true that a large part of the impact, both on the anthropic environment and the natural environment, is localised around industries, justifying the greater return of taxes to these places, but many factors such as the number of jobs generated and the area occupied are not taken into account in the distribution of the resources collected. What's more, environmental taxes, which are still almost non-existent, always end up in a common ditch, with no destination directed to the source.

On the physical side, companies in the sector have increased their environmental care due to the demands of the international market, but there is still a lot to be done. Legislation favours bureaucracy, while the technical aspect is due more to environmental certification programmes, which, fortunately, most of the large pulp companies have joined.

Income needs to be better distributed in this country and the resources collected need to be used more appropriately. The blame for the distortions lies not with the productive sector of the economy, but with misguided tax policies and environmental impact assessments.
Eucalyptus cultivation is an extremely controversial subject and, although I have my own opinion on many of the issues discussed below, it is not the intention of this work to dispute the negative or positive impacts attributed to eucalyptus cultivation, nor to offer alternative technologies and solutions, or even to compare current methods with technologies considered outdated or unfeasible by scientists and technicians in the field. Therefore, the main objective will be to survey, through a bibliographical review, all the impacts that may occur and identify the quantifications made within the current context and, based on this survey, try to subsidise the assessment of the environmental impacts caused, based on what has been discovered to date.

Maceió, 21st January 2024.

Eduardo Pagel Floriano

CHAPTER 1

INTRODUCTION

Brazil went from pulp importer to exporter in less than three decades. The policy of government incentives, represented by FISET-Reforestation and funding for the installation of forest-based industries with BNDES financing, which began in the mid-1960s, was the major impetus in this direction. According to PEREIRA (2001), FISET involved resources provided by tax incentives based on Law 5106/65, Decree-Law 1134/70 and Decree-Law 1376/74 which, according to JUVENAL & MATTOS (2002), totalled around 6 billion dollars by 2001, the year in which the forestry sector generated a GDP of 21 billion dollars, corresponding to 4% of the national GDP, and exports of 5.4 billion dollars, with a positive sector trade balance of 1.4 billion dollars.

Other factors that had a decisive influence on this were the discovery of eucalyptus by Brazilian foresters as a fast-growing and easily cultivated tree, the climate suitable for growing species of this genus and the availability of large tracts of land at low prices. From 500,000 hectares of economic forests planted in 1965, when the New Forest Code was enacted, marking the beginning of a new era, we went on to grow to around 4.8 million hectares of forests planted for industrial purposes in just over two decades with the tax incentives of the time.

According to BACHA (2000), in 1995, the wood agro-industrial system in Brazil (Wood SAG) was responsible for more than 3.26% of Brazil's GDP, producing 113.7 million m^3 of wood from planted forests and, in 1998, generating exports of US$ 3.65 billion and imports of US$ 1.14 billion; the Wood SAG is made up of all the segments offering products and services to forestry and forest extraction, forest harvesting itself, forestry, processing activities and distributors of products using wood and includes the following activities:

- industries that produce inputs to be used in plant extraction and forestry (such as tractors, equipment, pesticides and fertilisers), companies that sell these products, and service providers to forestry and plant extraction establishments;

- forestry and plant extraction establishments;
- industrial segments that carry out the first industrial processing of wood or charcoal, such as steel companies, mechanical wood processing companies (sawmills, veneer, plywood and fibreboard factories) and the pulp and paper sector;
- industrial segments that process products made from wood (furniture companies, printing and publishing companies, packaging companies, for example), or distribute the products of all the aforementioned segments.

In 2000, Brazilian short fibre pulp production reached around 5.5 million tonnes, according to BRACELPA (2001) *apud* SBS; which means the consumption of wood produced by around 620,000 hectares of *Eucalyptus* forests, considering a yield of 4.5 m3 of wood per tonne of pulp and forest productivity of 40 m3 of wood per hectare per year.

According to data from SBS (2003), in 2001 the Brazilian forestry sector had a GDP of US$ 21 billion, exports of US$ 5.4 billion, consumed 300 million cubic metres of wood (45 million m^3 of eucalyptus), collected US$ 2 billion in taxes and employed 2 million workers.

REVISTA DA MADEIRA (September 2001) reported that there were more than 13 million hectares of eucalyptus trees planted worldwide; in Brazil there were 2.2 million hectares, 45 per cent of which were used for the production of panels, paper and cellulose and 55 per cent for the production of firewood and charcoal: *Eucalyptus grandis* (55%), *Eucalyptus saligna* (17%), *Eucalyptus urophylla* (9%), *Eucalyptus viminalis* (2%), hybrids of *E. grandis* x *E. urophylla* (11%) and other species (6%).

The value of Brazilian forestry production in 2021, by state, is shown in Figure 1.

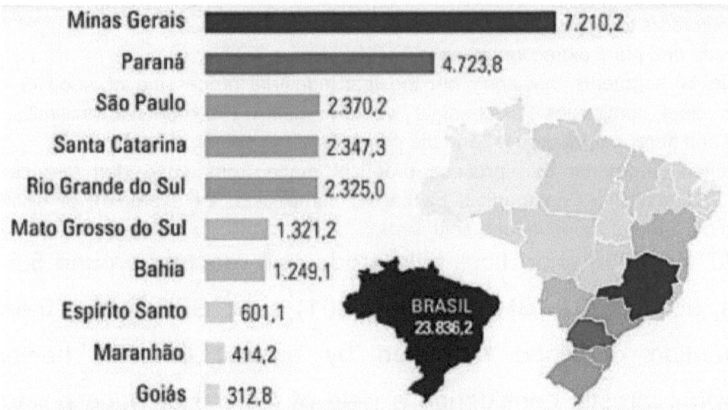

Figure 1 - Value of forestry production by state in 2021 (in millions of reais). Source: IBGE (2024).

In 2022, forestry GDP totalled R$27.4 billion. The value of production from planted forests surpassed that of plant extraction, which reached R$6.2 billion (IBGE, 2024), making the impact of extracting timber from natural forests less and less, due to the growing supply of forestry products to the market. The evolution of forest plantations in the country is shown in Figure 2.

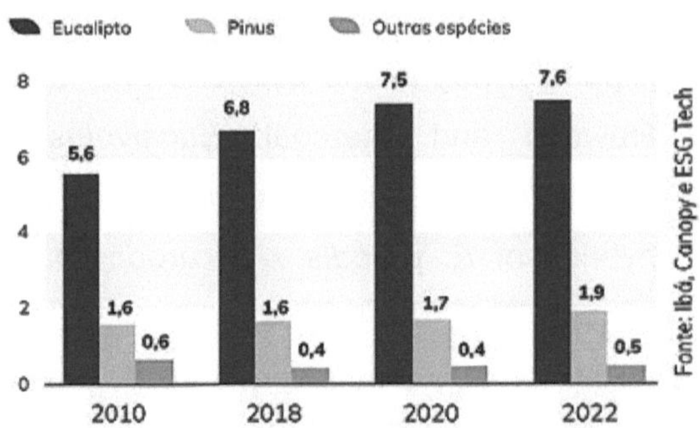

Figure - - Evolution of the area of planted forests in Brazil (million ha). Source: IBA (2023).

In 2022, planted forests totalled 9.94 million hectares, of which 76% were eucalyptus with 7.6 million hectares (IBA, 2023), used mainly in the pulp and steel industry.

Forestry activities result in various products, such as sawn timber, panels, pulp and fuel, the most important of which are the pulp production and steel segments.

Eucalyptus plantations by state are listed in Figure 3.

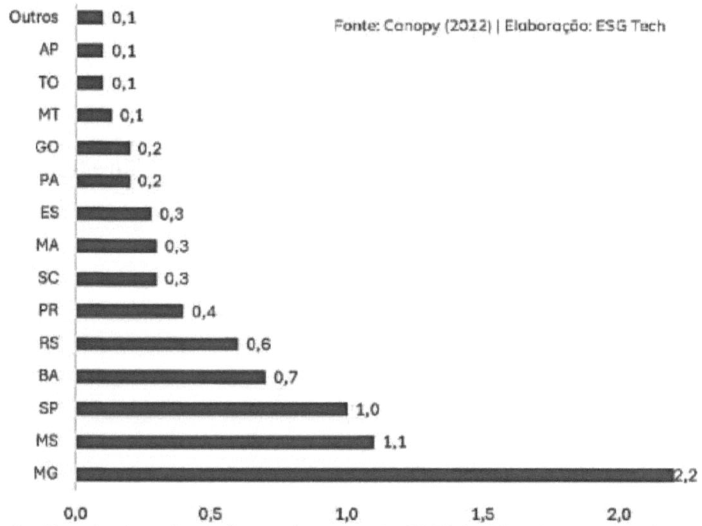

Figure 3 - Eucalyptus planted area by state in 2022 (million hectares). Source: IBÁ (2023).

Brazil's planted forests are concentrated in Minas Gerais, Paraná and São Paulo. In Minas Gerais, production is mainly destined for the steel industry; forests for the production of wood to be used in the manufacture of pulp and paper are mainly in São Paulo, Paraná, Bahia and Espírito Santo, where the two species most used for the manufacture of pulp, *E. grandis* and *E. urophylla*, have the highest productivity. In the southern region of Brazil there are large areas of *Pinus* production destined for the softwood pulp industry.

Some of the functions of planted forests are (SFB-SNIF, 2024):

> Reducing pressure on native forests;

> Reuse of degraded land;

> Carbon sequestration;

> Soil and water protection;

> Shorter rotation cycles compared to countries with a temperate climate;

> Greater product homogeneity, making it easier to adapt machines in

the industry.

The pulp and paper segment is responsible for a GDP of more than US$ 20 billion and more than 184,000 jobs and exports around US$ 8.4 billion. The entire planted forest sector generates almost 2 million direct jobs (Figure 4).

Although planted forests provide many benefits for human civilisation, there is a backlash against forest plantations.

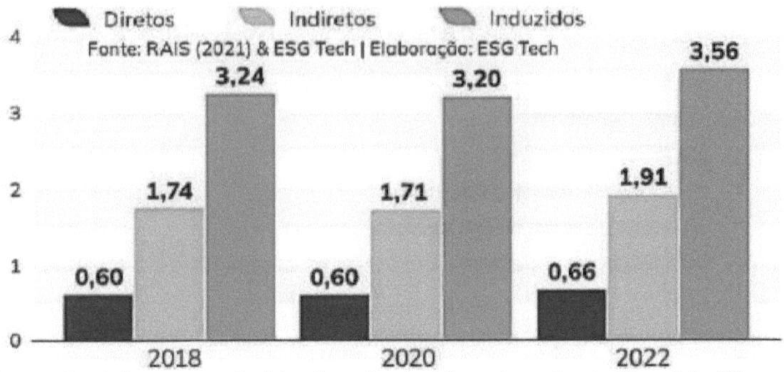

Figure 4 - Jobs generated by the planted forest sector in Brazil (millions of employees). Source: IBÁ (2023)

Eucalyptus species have been accused of many things, myths have been formed around their cultivation, but the truth is that eucalyptus farming generates both positive and negative impacts on anthropogenic and natural environments.

Examples of the environmental implications of eucalyptus forests include carbon sequestration by the planted forest, with a positive effect on the air's CO_2 content, reducing the greenhouse effect; the use of pesticides to control competition, pests and diseases, with negative effects on biodiversity; among the beneficial economic aspects is the contribution to Brazil's positive trade balance and among the negative ones, the foreign currency indebtedness of some companies; among the social aspects involved, the education and health programmes promoted by the pulp companies can be cited as positive examples, but the unemployment generated by the crisis in the sector around 2004 and the intense mechanisation of the sector was negative for society until it accommodated to the new situation.

One of the first major impacts of the sector to receive attention was the large consumption of forestry raw materials, which led the Brazilian government to adopt measures to guarantee the sustainability of companies in the production of the wood they consumed. This policy was structured by IBAMA's Normative Instruction No. 1 of 5 January 1996, which obliged large consumers of forestry raw materials to have an Integrated Forestry Plan (PIF). Forest replenishment was already provided for in the Forest Code of 15 September 1965, Federal Law 4771/65. Later, with the establishment of industries that began to consume wood from planted forests with the tax incentives of Law 5107/66 and Decree Law 1134/70, there was a need to guarantee the continuity of the forestry programme and the replacement requirement became one of the policies of the former IBDF - Brazilian Forestry Development Institute, which was later transformed into IBAMA, bringing together all the executive bodies in the environmental area at federal level.

The rules for forest replenishment and the consumption of forest raw materials were reformulated by MMA Normative Instruction No. 6 of 15 December 2006, but essentially the requirement for a Sustainable Forest Management Plan (SFMP) for natural forests and a Sustainable Supply Plan for the supply of forest raw materials to industries remains.

The environmental impacts of the pulp and paper industry began to disturb society, sometimes with good reason. Eucalyptus forests began to be viewed with suspicion, on the grounds that the species dries out the soil and depletes water sources, that nothing can be created under eucalyptus forests, that it depletes soil fertility; and when the first pulp mills using sulphur products were set up, the smell, the effect of the gases and the black stain of the effluents, which at that time had no filters, sparked movements such as the one promoted by agronomist

José Antônio Lutzenberger against the Norwegian company Borregard, in Rio Grande do Sul, then Klabin-Riocel, Aracruz Celulose and currently CMPC Brasil, which culminated in the closure of the mill from 7 December 1973 to 13 March 1974 and with the environmentalist taking over the Ministry of the Environment many years later, from March 1990 to April 1992.

As well as internal pressures, there are external ones, such as the visit to Aracruz in the early 1990s by England's Prince Charles, a shareholder in the company at the time, which led Greenpeace to move its ship to the Brazilian coast and anchor the boat, blocking the company's harbour during the visit, in protest at the environmental impacts generated by the pulp industry. Add to this the environmental control requirements of projects financed by international organisations such as the World Bank and the Inter-American Development Bank.

Also in the 1980s, Brazil created its environmental policy, expressed in Federal Law No. 6938/81, and industries such as Bahia Sul Celulose, built in 1989, were already born under the aegis of Environmental Impact Reports (RIMA) and adhered to environmental management programmes such as BS-7750[1] , in order to gain competitive advantages in the international market.

Environmental pressures have increased all over the world and in all activities, especially in the manufacturing industries. By the end of the 1980s, voluntary environmental control programmes were already in place in many companies. In the 1990s, the evolution of international environmental standards led to the ISO 14000 series. There were also studies on chlorine and dioxins, with European pressures against products manufactured using chlorine, which led all pulp companies to change their chlorine bleaching processes to processes using oxygen.

Finally, we come to the Brazilian legal requirements for setting up

[1] Note: British environmental management standard, precursor to the ISO 14000 series.

forestry projects, which include environmental impact studies and assessments. Environmental impact assessments in this sector of the economy also serve companies' environmental management programmes, mainly with the intention of certification by international bodies, which has represented a quality differential for marketing pulp and paper on a global level.

Many methodologies for environmental impact assessment have been suggested. There is still no complete standardisation in this regard and each case is studied individually by the official licensing bodies and certifying organisations, but the paths towards this are taking shape. Some environmental quality indicators have limits that are required, while others are negotiated. The Forest Stewardship Council (FSC) and International Organisation for Standardisation (ISO) 14000 series standards represent a considerable step forward and, as well as guiding the creation of environmental management systems, allow them to be adapted to any case, in addition to requiring compliance with all related legislation. The forestry sector is also covered by a sub-series of ISO 14000, due to the major impact of forests on the global environment. The main aim of the two organisations' forest certification programmes, among others like them around the world, is to encourage companies to create forest management programmes that are environmentally appropriate, economically viable and socially just. According to MALINOVSKI and CAMARGO (2001), eucalyptus cultivation is highly promising in Brazil, but ISO 14001 or FSC certification will be mandatory in all markets.

The aim of this study is therefore to characterise Brazil's production of *Eucalyptus* wood for pulp production and to survey its environmental consequences in order to help prevent negative aspects caused by the crop.

CHAPTER 2

2 CHARACTERISTICS OF THE *Eucalyptus* WOOD INDUSTRY

2.1 PULP AND PAPER SECTOR

According to SOUZA and OLIVEIRA (2002), the Brazilian pulp and paper industry specialises mainly in the production and export of bleached, short-fibre sulphate (Kraft) pulp, which comes from eucalyptus as the main raw material. As verified directly with a significant number of companies in terms of representation in total Brazilian production, *Eucalyptus* pulp is produced by large companies that own large areas of coppice managed forests, with an average cutting cycle of 7 years and two to three harvests in the same area per rotation, which varies from 14 to 21 years. These are companies that use high technology, from seedling production to planting and maintaining forest stands, with a high degree of mechanisation in all operations, from seedling production to harvesting carried out by computerised machines. Wood is transported to the mills by heavy and extra-heavy lorries and the industries are highly automated, using a wide range of chemical products. Eucalyptus pulp is mainly used to produce printing, writing and hygiene paper, which end up in consumer products with a certain degree of permanence in use, reusability, recyclability and degradability after disposal.

SOLITRENICK *apud* BRACELPA, in his booklet "*History of Cellulose*", records the beginning of the activity in Brazil:

> Large-scale production of Eucalyptus pulp using the Kraft process began in the state of São Paulo in 1957, paving the way for the great pulp industrialisation stage which, in less than a quarter of a century, took Brazil to the position of 8th largest producer in the world.

According to BRACELPA, in 2000 sales of short fibre pulp reached more than 5 million tonnes and according to available data, it is estimated that production in 2003 should exceed 8 million tonnes. Brazilian pulp production in 2001 and in the first half of 2002 (Table 1), with bleached hardwood pulp, produced almost exclusively from eucalyptus wood, accounting for more than 70% of the total.

In 2022, pulp production in Brazil reached 25 million tonnes, of which 22 million tonnes were short fibre (more than 4 times what was produced in 2000), 2.5 million tonnes were long fibre and 0.5 million tonnes were high-yield pulp (IBÁ, 2023).

The main applications of the different types of cellulose produced in the country and their uses are as follows:

> Bleached eucalyptus pulp - printing and writing paper, sanitary paper, manufacture of chemical (self-copying) or thermal (fax and commercial automation) paper, cardboard;

> Unbleached eucalyptus cellulose - decorative laminates, formica, filters and cardboard coatings for packaging and carbon base;

> Bleached sisal cellulose - cigarette papers, filters, sachets;

> Unbleached sisal cellulose - dielectric papers, fibre cement boards;

> Abaca pulp - high porosity papers;

> Fluff cellulose - nappies and pads.

TABLE 1 - Brazilian cellulose pulp production (tonnes)

PRODUCT	2001	PERIOD JAN-JUL/2001	JAN-JUL/2002
Long fibre	1.438.495	816.780	877.778
. Bleached	70.985	43.463	47.263
. Unbleached	1.367.510	773.317	830.515
Short fibre	5.504.971	3.146.573	3.358.235
. Bleached	*5.292.351*	*3.029.399*	*3.215.060*
. Unbleached	212.620	117.174	143.175
High performance pulp	468.561	276.129	274.338
TOTAL	**7.412.027**	**4.239.482**	**4.510.351**

Source: BRACELPA, 2002.

Brazil's *per capita* paper consumption of around 44kg *per* year in 2022 indicates great potential for the domestic market, as the average in developed countries is much higher. The foreign market shows a continuous increase in consumption, with no tendency to change (Figure 5).

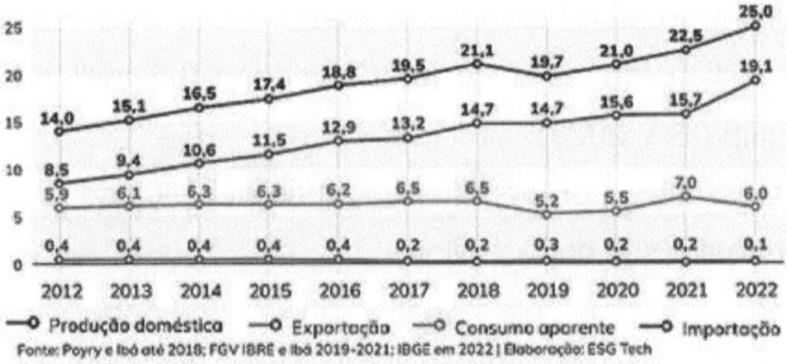

Figure 5 - Pulp production and consumption in Brazil. Source (IBÁ, 2023).

The overall consumption of paper shown in Figure 6 demonstrates the small consumption in Latin America; in 1994 per capita consumption in Brazil was less than 10 per cent of the US at 332 kg/year/inhabitant (MACEDO E VALENÇA, 1994).

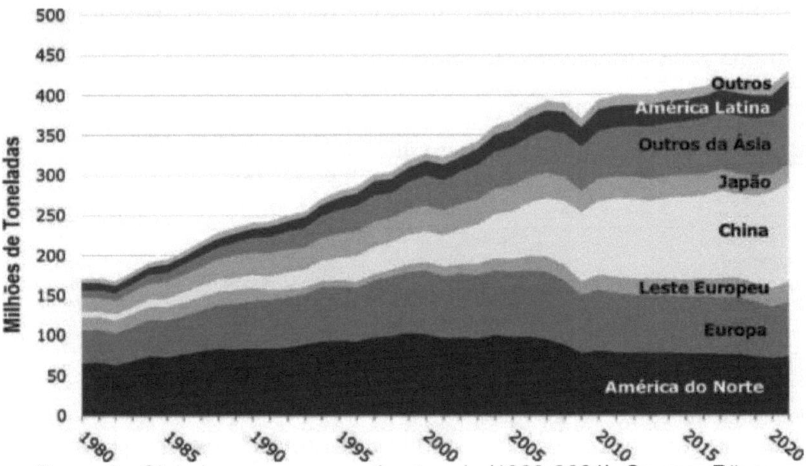

Figure 6 - Global paper consumption trends (1980-2021). Source: Pöyry *apud* Neves and Porto (2023).

Table 2 compares the performance of Brazil and the USA in the pulp and paper sector in 2004.

TABLE 2 - Pulp and paper overview: Brazil vs USA * (2004)

PARAMETERS	BRAZIL	USA
Commercial forestry area:	1,415,936 km^2**	1,955,965 km^2*
Pulp and pulp capacity:	8,338,000 tonnes*	93,040,000 tonnes*
Pulp capacity:	8,075,000 tonnes*	62,986,000 tonnes*
***Per capita* Pulp and Pulp Cons:**	**38kg***	**324 kg***
No. of employees:	100.000*	198.800*

Sources: (*) PPI; (**) BRACELPA; (***) Estimate.

Figure 6 shows the evolution of world market pulp consumption from

1990 to 2021.

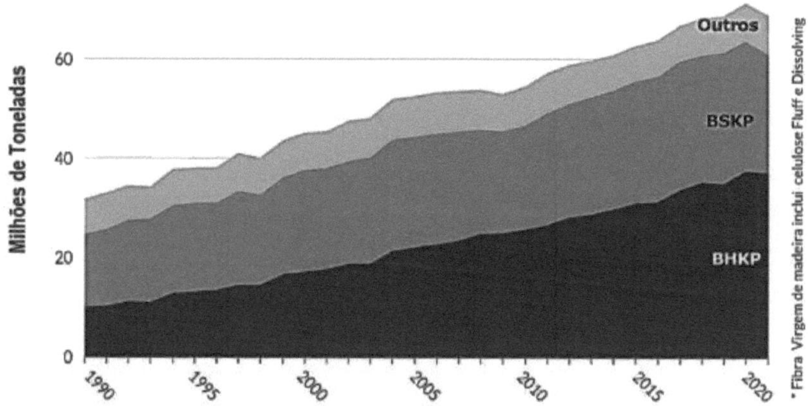

Figure - Global consumption of market wood pulp by type (19902021). Source: Pöyry *apud* Neves and Porto (2023).

Although Brazil and the USA are completely different, the situation in the USA in a way represented our growth trend with a few caveats, for example in terms of pulp volume, which in the USA represented around 1/3 of production, in the Brazilian case it was insignificant, mainly due to the species used, conifers and broadleaves respectively, as a reflection of the difference in climate.

The largest pulp producers are listed in Figure 8.

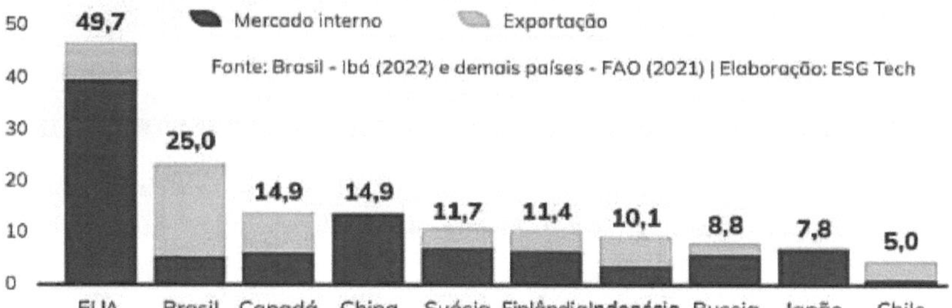

Figure 8 - Top pulp producers in 2021 (million tonnes).
Source: IBÁ (2023).

Pulp prices, which have been above US$ 800 per tonne during periods of scarcity and below US$ 300 during periods of excess stocks worldwide, have become more stable every day, remaining between US$ 400 and 600 per tonne of bleached softwood pulp between 1996 and 1999 (Figure 9); but, like any *commodity*, they can show greater variations if stocks are not under control. Pulp prices are mainly affected by stocks in the NORSCAN system.

Short fibre pulp prices have risen to US$ 610 a tonne for China for 2024 (INFOMONEY, 2023).

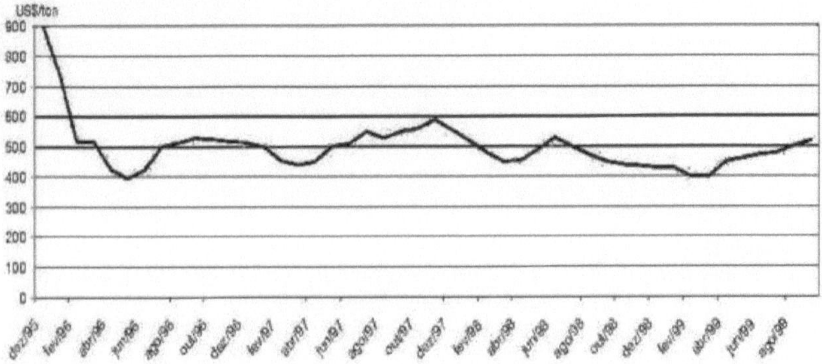

Figure 9 - Variation in bleached eucalyptus pulp prices, CARVALHO (2000).

Prices for market pulp fluctuate widely, causing crises in the sector, including one in the early 1990s, which led many companies to go bankrupt, especially in northern Europe. In Brazil, according to the BNDES (2002), companies are better able to withstand price fluctuations due to the low costs and high productivity of *Eucalyptus* forests, with advantages for large-scale pulp production that are recognised worldwide. World paper production and consumption have been growing for more than fifteen years. The Bank also reports that the growth rate in world demand for pulp and market pulp between 1990 and 2000 was 3.5 per cent per year, with demand for eucalyptus pulp growing proportionally more than for other types.

In October 2023, pulp prices in Europe reached an average of 980 dollars a tonne, while in North America they reached 1170 dollars (UOL-ECONOMIA/REUTERS, 2023).

There is a trend towards new investments in Brazil, mainly because the industrial park of some of the largest producers in the northern hemisphere is getting older every day and is based on low-production forests. Our industries are recent, some of them state-of-the-art, polluting little in relation to the world average, and our forests have high and growing productivity. The Brazilian pulp and paper industry is gaining a reputation for the quality of its products and for its environmental care; with this it is gaining market share and attracting foreign investors with a tradition in the sector, such as International Paper and Bracell. Against this backdrop, there were expectations of an increase in Brazilian production of over 35 per cent for the five years following 2004, a considerable increase in environmental care and investments in technology with a view to increasing the quality and productivity of both forests and industries.

The productivity of *Eucalyptus* forests in the early 2000s, according to information from some companies, was: Jarí 37 m^3 of wood/ha/year; Cenibra 38 m^3/ha/year; Bahia Sul had a programme to reach 45 im3/ha/year, a figure that Aracruz was already exceeding in its plantations in Bahia and Espírito Santo. Klabin-Riocel even sold part of its eucalyptus plantations in Rio Grande do Sul due to the increase in productivity[2] . One company recently announced productivity of up to 60 m^3 /ha/year in some stands.

The sector's companies have invested heavily in research, both operational and genetic and silvicultural, with a view to increasing the productivity of plantations and forestry operations. It is estimated that the sector's average forestry productivity was around 40 m^3 of wood per hectare at the end of the 20th century (REVISTA DA MADEIRA, September 2001).

In the operational area, forest monitoring and mechanisation have been increasing productivity rates. While monitoring plantations reduces failures, prevents the spread of pests, diseases and fires and identifies

[2] Klabin sold Riocel to Aracruz in May 2003; the former Riocel now belongs to CMPC, which modernised and expanded it.

nutritional deficiencies, large harvesters have replaced chainsaws with exceptional operational gains in productivity and, in all operations, methods, machines and tools are being introduced that have profoundly increased operational productivity rates, such as minimum cultivation, planters and furrowers. As a result, Brazil has become more competitive on the international market, with average costs of US$169 per tonne of pulp, a figure only beaten by Indonesia (Figure 10), according to International Paper, the world's largest group in the sector in 2001, with revenues of over US$26 billion.

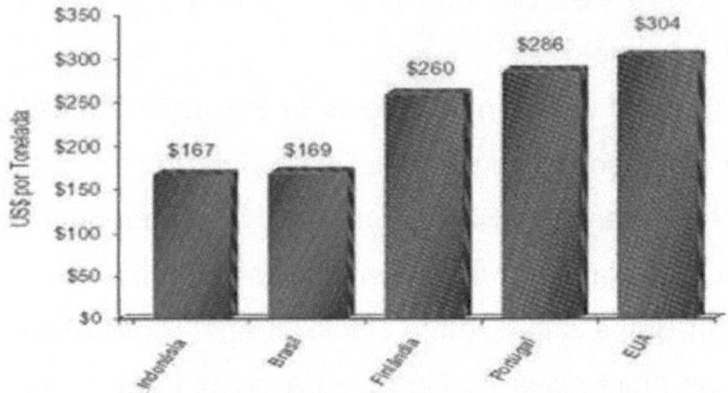

Figure 10- Pulp production costs, VCP (2003) - adapted.

The cubic metre of eucalyptus wood went from R$40.20 in 2019 to R$115.40 in 2023 (JORNAL OPÇÃO, 2023), with a strong impact on the sector's economy.

With increasing productivity and mechanisation in the forestry sector, the outlook for an increase in the number of jobs is not great. Mechanisation is an irreversible process due to the need for accelerated cost reduction and professionalisation in the sector in order to remain competitive at international level.

Aracruz is considered to be one of the companies that has mechanised forestry operations the most, with ideal conditions for this, having ended 2000 with a production of 1,301,000 tonnes of pulp[4] and with 1,585 employees, including the international subsidiaries and Aracruz Produtos

[4] Aracruz's 2002 report indicates a production of 1,656,000 tonnes.

de Madeira; adding to this the 2,954 outsourced employees, this makes a total of 4.539 direct jobs generated by the company, which results in 3.48 direct jobs for every 1000 tonnes produced, according to the company's annual reports; extrapolating these figures to the approximately 5.5 million tonnes of hardwood pulp produced by Brazil in 2000, there would be a total of just over 19,000 jobs, which can be compared to the figure of 30,331 jobs in Bracelpa's statistics for the same year. Aracruz generated around 50,000 indirect jobs. If we take the above figure, there would be 38.43 indirect jobs per 1,000 tonnes of pulp per year, or approximately 211,000 indirect jobs in the activity. It should be noted that the company worked in mostly flat and well-drained areas, facilitating the mechanisation of forestry operations, had a genetic base of extremely high productivity and operated in soil and climate conditions that provide excellent forestry yields.

According to the Ministry of Development, Industry and Foreign Trade (BRASIL, 2003), the Brazilian pulp and paper sector was made up of a group of 220 companies that directly employed 102,000 people, 64,000 of them in industry and 38,000 in their forestry activities.

In the social area, the Brazilian wood and pulp segment is atypical, as it invests heavily in many social projects due to its special characteristics, both because of its economic success and the environment in which it operates. The segment's social role was described by TABAKOF (2001), President of BRACELPA, when he presented the sector's social report:

This Social Report focuses on the men and women who work for Brazilian pulp and paper companies, as well as the communities in which we operate. This publication is based on data obtained from a
sample of 36 Bracelpa member companies, which account for more than 90 per cent of the sector's production volume.

Equipment, machinery, large factories, cultivated forests and export figures usually make up the first image that comes to mind when we think of our sector, characterised by its industrial and agricultural outlines. Our sector's large size and long history of pioneering spirit and entrepreneurial capacity fully justify this image.

But the significant social contribution of our industry is not clearly demonstrated by this image. In 2001, the pulp and paper sector collected R$1.840 billion in taxes, provided

130,000 direct and indirect jobs throughout Brazil and paid industrial wages to its employees totalling R$990 million.

In Brazil, unlike in more developed countries, private companies also play an important role in inducing social development, in addition to playing an important role in economic growth. This social role is even more pronounced in the pulp and paper agroindustry, because our industrial facilities and forestry areas are located far from urban centres. For this reason, our companies need to make substantial investments in favour of the social well-being and infrastructure of the communities where we operate.

Only the investments made in this field during 2001 are reflected in the figures contained in this Social Report. It should therefore be remembered that the substantial investments that have been made for decades by our companies in this field are not included here. These investments, in addition to giving the sector high social relevance, have given Brazil an industrial position of global prominence, which places it in 11th place among the world's paper manufacturers and in 7th place as a pulp producer.

The self-sustainability that characterises our industry is one of the main factors that has ensured our country this honourable position at the forefront of world manufacturers. The companies that make up the sector have always been dedicated and serious about industrial and forestry environmental protection: as well as being equally concerned with the preservation and cultivation of native forests, they use only eucalyptus and pine planted for this purpose - natural and renewable resources - as raw materials for their factories.

It is for all these reasons that we feel honoured to present to readers in Brazil and abroad this Social Report - 2001, in which we bring together the various investments made for the benefit of our country's population by the pulp and paper companies represented by Bracelpa.

Some of the activities in which pulp and paper companies have invested that result in benefits for society are:

> building and equipping hospitals, health centres, nurseries, schools, sports centres and vocational training centres;

> building, equipping and maintaining centres for the development of phytotherapy and the cultivation of medicinal plants;

> construction and financing of homes, in some cases with the construction of residential neighbourhoods with all the infrastructure[5] such as supermarket, bank, school, social club, telephone exchange, etc;

> promotion and sponsorship of cultural and sporting events;

> building and improving public leisure areas;

> building and improving local roads for their own use and that of third parties;

> building and equipping airfields;

[5] This includes Klabin in PR, Aracruz in ES, Bahia Sul in ES and BA and Jarí in PA.

> support and donation of machinery, equipment and supplies to town halls;

> support in preventing and fighting rural and urban fires outside their areas;

> alcohol and drug recovery programmes for the society in which they operate;

> road accident prevention programmes;

> programmes to collect and distribute warm clothes and food;

> supporting and equipping official public security services;

> in the environmental area: implementation of extensive environmental education programmes; support for scientific wild animal breeding; creation of Private Natural Heritage Reserves (RPPN); creation of ecological trails; sponsorship and support for environmental research projects; creation and maintenance of environmental and forestry research and development centres; monitoring of river basins; creation and maintenance of herbaria and flora and fauna museums.

According to SILVEIRA *et al.* (2001), the Wood SAG, made up of the pulp and paper, charcoal, sawn timber, board and chipboard industries, has been making a name for itself in the national economy, especially in the pulp and paper segment where, in the 1950s, Brazilian pulp production was a meagre 95,000 tonnes, rising to 6.3 million tonnes in 1997.

Due to the competition and demands of the international market in the 1990s, practically all the big companies in the sector went for certification according to the International Organisation for Standardisation (ISO) 9000 series standards, and some also the ISO 14000 series, or sought the Forest Stewardship Council (FSC) green seal as a way of guaranteeing international standards for their products for export.

Among the nine largest Brazilian companies in the sector in 2001, Aracruz stood out as the largest producer of wood pulp.

Eucalyptus and Klabin as the largest pulp producer in general[5] (Table 3).

Currently, the largest producer of eucalyptus pulp (short fibre) is Suzano and Klabin continues to be the leader in the production of long fibre pulp.

TABLE 3 - The 9 largest pulp producing companies in Brazil in 2001

COMPANY	PRODUCTION CELLULOSE [b] (THOUSAND TON.)	AREA (Thousand HA)			
		PRODUCT FORESTS [c]	PRESERVATION and Legal Reserves	OTHER	TOTAL
Aracruz	1585	144	65	13	222
Bahia Sul	578	76	46	8	130
Cenibra	792	115	85	32	232
IP	498	207	265	128	600
Jari	330	45	1480 [a]	209	1734
Klabin	2100 [d]	231	135 [a]	56	422
Ripasa	310	57	18	2	77
Suzano	420	80	46 [a]	27	153
VCP	800	102 [a]	47 [a]	8 [a]	157
SOMA	7413	1057	2187	483	3727

Notes on the figures presented: (a) estimated based on information on the company's website; (b) pulp production includes all types, including softwood; (c) the area of production forests includes all cultivated species of Pinus, Araucaria, Eucalyptus and others. (d) Klabin - total produced including paper, pulp and paper products - information gathered from the group's website on all its companies (including Riocel, which was sold to Aracruz in May 2003). Aracruz, Bahia Sul, Ripasa and VCP were acquired by Suzano. Source: Company websites.

According to data collected from the company's websites and the Bracelpa, investments in the eucalyptus pulp sector were planned for the years following 2004, which should make some of the big companies even bigger, and new companies would join the ranks of the big :

> Veracel, a company formed by the association of ARACRUZ (45%), Stora Enso (45%) and Odebrecht (10%), has plans to set up a new plant in the south of Bahia; it will have the capacity to produce 900,000 tonnes a year and will require US$ 1.25 billion in investments, of which US$ 300 million have already been made in the forestry area; it will need approximately 90,000 hectares of eucalyptus forests;

> Lwarcel, part of the LWART group, with a small mill and 25,000 hectares of forests, plans to produce 180,000 tonnes of market pulp a year;

> VCP plans to increase its production from 800,000 to 1,200,000 tonnes of pulp per year in 2003;

> Klabin increased production at the Guaíba mill (Riocel) to 400,000 tonnes of pulp in 2002/3, a one-third increase on 2001 capacity; Riocel was sold to Aracruz in May 2003, which promised to modernise the

company and make investments to make it more competitive;

> Celmar, in Maranhão, planned to enter the market with an annual production of 610,000 tonnes of pulp, requiring around 70,000 hectares of productive forests; its forestry activities began in the mid-1990s.

Table 4 shows the evolution of the domestic pulp and paper market from 1996 to 1999.

Brazilian pulp production is considerable for the country's economy, showing a strong and constant growth trend, indicating that it is an economic sector that deserves special attention due to the impacts it generates in all areas, both positive and negative.

TABLE 4 - National pulp and paper market (thousand tonnes)

PARAMETERS	1996	1997	1998	1999
Total Brazilian Production	6.201	6.331	6.800	7.209
Bleached Short Fibre Production	4.098	4.333	4.800	5.091
Export	2.400	2.344	2.450	3.012
Pulp Mills	34	33	33	100
Short Fibre Pulp Mills	17	18	18	13
Total Brazilian Production	6.199	6.518	6.580	6.889
Export	1.400	1.280	1.220	1.257
Printing and Writing Paper	1.806	1.969	1.990	2.070
Domestic Sales	1.071	1.073	1.130	1.517
Export	690	715	680	750
UWF - Production	1.500	1.650	1.660	1.672
Export	600	1.650	1.660	663
Paper Mills	186	178	178	181
Printing and Writing Paper Mills	25	25	25	19

Source: BRACELPA *apud* INTERNATIONAL PAPER. Note: The total number of factories refers to the companies participating in BRACELPA's statistics.

2.2 LAMINATE FLOORING AND PANELLING SECTOR

The domestic wood panelling market in 2022 reached sales of 7.0 million m^3. MDF panels accounted for 61% of sales, or a total of 4.2 million m3. Brazil maintained its 8^a position in the global ranking of wood panel production, with 8.5 million m3 produced (IBÁ, 2023).

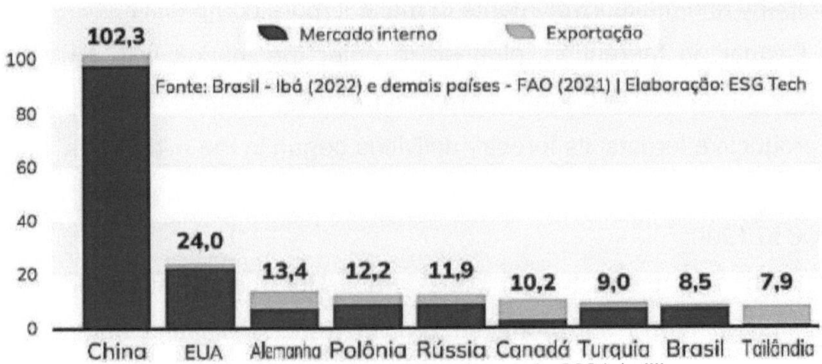

Figure 11 - Top wood panel producers in 2021 (million m3). Source: IBÁ (2023).

Wood panelling can be made from different types of resin, resulting in products with different applications, such as:

> Resins based on urea formaldehyde - used for gluing furniture plywood, joinery, veneer joints and finger joints, whose products are used indoors as they are not water resistant;

> Melamine Urea Formaldehyde Resin - used for hot press bonding, where the material produced needs to be more resistant to water and humidity;

> Melamine resin - used for bonding products for the international market that does not allow the use of phenolic resins. The resistance of this resin is similar to that of phenolic resin.

> Phenolic Resin - used for bonding wood veneers in general in hot processes for products that require high water resistance.

The main types of wood panelling are

> Plywood - is made from thin sheets of pressed wood, glued together, interlaced orthogonally to each other in relation to the wood fibres; it is resistant and durable; used in furniture making, construction and shipbuilding;

> OSB (Oriented Strand Board) - a panel made up of strips of wood arranged in the same direction; used for siding, internal walls and rustic-looking furniture.

> MDF (Medium Density Fiberboard) - medium-density fibreboard, made by bonding wood fibres with synthetic resins and additives; the panel is solid and consists of a single homogeneous layer, giving it a smooth surface; it can withstand machining such as bending, notching and milling;

> MDP (Medium Density Particleboard) - medium density particleboard, made from wood particles in three layers; cannot be machined;

> HDF (High Density Fiberboard) - high-density fibreboard; made from wood fibres thermo-set by pressure and high temperature and impregnated with resin, guaranteeing strength, a uniform surface and a smooth texture; used in the bottom of furniture and drawers, curved pieces and niches; can withstand carving and milling.

Laminate flooring sales in 2022 totalled 10.5 million m^2, a reduction of 23.7% compared to 2021, with around 95% destined for the domestic market (IBÁ, 2023).

Laminate floors are boards (or strips) made of wood fibres or particles, reproducing woody patterns on the surface, with high resistance to abrasion and consequent durability.

According to ABAF (2024), the composition of laminate flooring is as follows:

"**Surface Layer (Cover)**

Overlay - A crystalline cellulose film impregnated with melamine resin with aluminium oxide particles, responsible for its high resistance to scratches and abrasion. It stands out for its low porosity, thus preventing the absorption of liquids and making it easier to clean, as well as providing durability.

Decorative veneer - Decorative veneer made from cellulose, impregnated with melamine resin that adds beauty to the floor, providing a variety of patterns and combinations.

Substrate (Core)

HDF / HPP - High-density wood fibre or particle board, produced from planted pine and eucalyptus trees in accordance with global environmental standards, and machined on its sides and top to establish tongue and groove joints. The substrate provides robustness, stability, precision fitting between the boards and resistance to the impact and concentrated load of heavy furniture.

Balancing Layer

Balance or Underlay - Made of cellulose sheets impregnated with melamine resin, which gives the floor stability."

2.3 CHARCOAL SECTOR

Brazil leads the global ranking of charcoal producers, reaching the mark of 7.0 million tonnes produced in 2022, with almost all production going to the domestic market (Figure 12).

Of this total, 6.9 million tonnes of charcoal were obtained from wood from planted forests, which represents an increase of 15% over the last 5 years. According to information from SINDIFER, the Steel and Steelmaking sector produced 7.8 million tonnes of pig iron from charcoal in 2022, totalling 24.1% of its total production (32.4 million) from charcoal.

The specific consumption of charcoal is 2.9 m^3 / t of pig iron (5) or 0.725 t of charcoal / t of pig iron, so the mass of pig iron produced per tonne of hanged wood is 0.23 t. The typical carbon content of pig iron is 4.3% by mass, meaning that 1 t of pig iron requires 0.725 t of charcoal, produced from 3.6 t of wood (FERREIRA, 2000).

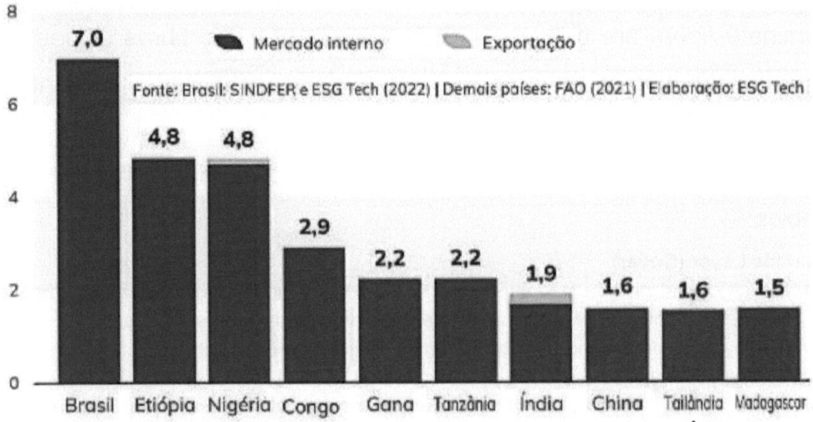

Figure 12 - Main charcoal producers (million tonnes). Source: IBÁ (2023).

The evolution of charcoal consumption in the steel industry is shown in Figure 13.

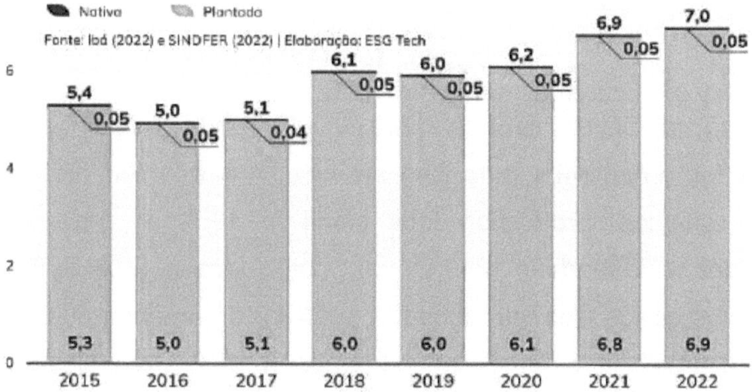

Figura 13 - Consumo de carvão vegetal na siderurgia (milhões de toneladas).
Fonte: IBÁ (2023)

CHAPTER 3

3 *eucalyptus* forestry in brazil

3.1 History of eucalyptus cultivation in Brazil

According to SCARPINELLA (2002), eucalyptus was introduced in Europe for ornamental purposes around 1774 and the first trials for commercial-scale production took place in 1854; in India, the first plantations were made in 1843 and in South Africa in 1828; In Brazil, the *Eucalyptus* genus was introduced in 1825 for ornamental purposes and as a windbreak. Its commercial use began in 1903, when sleepers were needed to meet the development of the railways through the Companhia Paulista de Estradas de Ferro; its use at the time extended to the production of firewood and charcoal. According to GALETI (1973), Edmundo Navarro de Andrade, head of the forestry service at Companhia Paulista de Estradas de Ferro from 1903 to 1941, brought 125 species of *eucalyptus* to Brazil in 1909; according to RIZZINI (1971), he cultivated 144 species of *Eucalyptus* in São Paulo, having started the activity in 1904 (REVISTA DA MADEIRA, Sept./2001). Navarro de Andrade is regarded as the father of eucalyptus cultivation in Brazil.

The genus *Eucalyptus*, with more than 600 identified species, belongs to the family *Myrtaceae* L'Her., subfamily *Leptospermoideae*; it is native to Australia, Tasmania and Indonesia (Java, Papua, Philippines and the Sonda archipelago: Timor, Flores, Adonara, Lomblem, Pantar, Alor and Wetar), with a latitudinal variation from 7° North to 43° 39' South; it has species ranging in size from shrubs to trees 100 metres high. Among the species most cultivated in Brazil for pulp, *E. urophylla* is native to some eastern islands of the Sonda archipelago and *E. grandis* and *E. saligna* come from eastern Australia (SCARPINELLA, 2002; LIMA, 1993; RIZZINI, 1971).

The most planted species, due to their higher productivity in the main growing regions in Brazil and the quality of the wood for pulp production, are *E. saligna* (17%), *E. urophylla* (9%), *E. grandis* (55%), according to

SCARPINELLA (2002). *grandis* (55%), according to SCARPINELLA (2002); but many hybrids of the latter two species are used with each other and with others, which are not included in the percentages mentioned, with the *Urograndis* from Aracruz cited by ZOBEL & *TALBERT* (1984) being the most famous crossbreed and first reported as a potential productivity flagship for a company, back in the 1980s. Natural hybrid crosses are common among *Eucalyptus* species. According to REVISTA DA MADEIRA (2002), Brazil has the world's highest average productivity in eucalyptus cultivation for pulp, 40 m^3 per hectare per year.

Due to the diversity of species, the genus has broad edaphoclimatic adaptation, with species that can adapt to dry to rainy climates ranging from -18°C to +35°C and from sandy to waterlogged soils (SCARPINELLA, 2002). The *E. dunnii* species grows reasonably well in climates prone to frost. The average productivity of eucalyptus in Brazil in 2022 was 32.7 m3/ha/year according to IBÁ (2023).

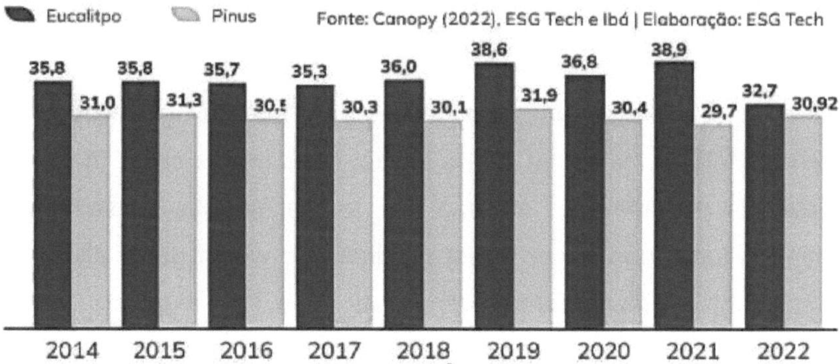

Figure 14 - Forest productivity (m^3/ha/year with bark). Source: IBÁ (2023).

Eucalyptus wood has a wide range of uses and the management system depends on the future use of the wood. The coppice silvicultural system is generally adopted, as the plants are easy to regrow, with cycles ranging from 5 to 25 years. Short cycles, with shallow cutting and in a coppice system, are used to produce wood for pulp, paper, charcoal, firewood, pallets, staking and joinery. The wood is usually harvested between 7 and 10 years after planting, allowing up to three regrowths, but rarely more than one regrowth due to loss of productivity. Long cycles at regular height and intermediate thinnings are used to produce sawn timber or veneer for furniture and construction.

3.2 Forestry phases, activities and operations

For the production of wood for pulp production, eucalyptus is grown in a coppice system with a 7-year cutting cycle and a regrowth with subsequent reformation and includes the following silvicultural phases (Table 5):

TABLE 5 - Short-cycle eucalyptus farming activities

	ROTATION					
	1ª		2a		3a	
	CICLOS		CICLOS		CI	CLOS
CYCLE YEARS	1°	2°	1°	2°	1°	2°
1	Planting	Conducting regrowth	Reform	Conducting regrowth	Reform	Conducting regrowth
2	Maintenance	Maintenance	Maintenance	Maintenance	Maintenance	Maintenance
3	Maintenance	Maintenance	Maintenance	Maintenance	Maintenance	Maintenance
4	Costs	Costs	Costs	Costs	Costs	Costs
5	Costs	Costs	Costs	Costs	Costs	Costs
6	Costs	Costs	Costs	Costs	Costs	Costs
7	Harvesting	Harvesting	Harvesting	Harvesting	Harvesting	Harvesting
	1ª Phase Implementation	2nd Phase Rebroken	Stage 3 1st Reform	4th Phase Rebroken	Stage 5 2nd Reform	6th Phase Rebroken

These phases and activities are typical of *Eucalyptus* plantations for the production of wood for pulp production, with the aim of maximising the technical yield of the forest at the lowest cost. In this case, the second regrowth is generally not allowed due to the mortality of stumps and planting failures and lower forest productivity, which generally justifies the costs of reforming forest stands after the first regrowth. In some cases, due to excessive planting failures before harvesting and low productivity, even the first regrowth of stumps is not allowed.

The silvicultural phases begin with the planting of the forest, which undergoes one or two regrowths, then the forest is reformed (replanted), with the remaining phases being a repetition of regrowth and reformation, indefinitely.

During all phases, monitoring and control of planted areas and forest growth (inventories), operational activities, pests and diseases, soils, climate, fires, hydrology, flora, fauna and infrastructure are carried out, both for quality assurance and for forestry and environmental management.

The operations in each phase are described below, based on technical visits to the companies: Ripasa (1991), Jarí (1992) and Suzano (1992); services provided to the companies: Cia Vale do Rio Doce (1985/86), Florestas Rio Doce S.A.[6] (1994 and 1998), Cenibra (1995 to 1998) and Aracruz (1994 to 1998); and as an employee of the companies: Florestas Rio Doce S.A. (1987 to 1989) and Bahia Sul Celulose S.A. (1989 to 1993). Data from 1998 onwards was updated using specialised bibliography, information made available on the internet by the organisations involved and through the companies' annual reports published on the internet.

3.2.1 Pre-silvicultural phase:

Before starting phase 1ª, forestry implementation, a number of activities have to be carried out, including activities that are not typically forestry, such as land acquisition and the construction of infrastructure for administration, access and security of the areas intended for cultivation, which also cause impacts of all kinds.

> Land acquisition: an operation that involves large sums of money and local inflation of land values, migration of people from the countryside to the cities or to other regions; it is carried out at least 10 years before the construction of a new factory or the expansion of old ones; part of the indigenous rural labour force is used in the new plantations; in many cases there is an import of labour due to local shortages, both in terms of quantity and qualifications.

> Construction of infrastructure: this includes the opening, construction and improvement of roads and firebreaks; construction of fences;

[6] Note: Incorporated into Bahia Sul in 1999.

construction of administrative headquarters, input depots and workshops, timber depots and warehouses, installation of fire prevention towers and communication systems. In some cases, the entire infrastructure of a town was built, in addition to the operational infrastructure, as in the case of Aracruz, Klabin, Jarí and Bahia Sul.

3.2.2 Silvicultural operations in phase 1ª - Planting:

> Seedling production: is carried out in tubes in nurseries, planting seeds, cuttings or micro cuttings (clones); the tubes are packed in trays that are installed in vegetation houses for germination or rooting for around 30 days and then in open areas for rustification for around 60 days, when the cuttings are then taken to the field for planting around 90 days after planting in the tube. This includes the production of seeds in seed production areas (APS), or the production of cuttings in sprout production areas, which is sometimes a conventional planting area used for this, or a clonal garden; *in vitro* reproduction is rarely used commercially; preparation of cuttings; sanitisation of packaging; preparation of substrate; preparation of the rooting inducer, usually AIB[7]; micro cuttings from clonal mini-gardens do not need AIB to root; rooting in the greenhouse; rustification of the cuttings in an acclimatisation area; fertilisation, irrigation and phytosanitary treatments; sanitation and storage area for materials and tools.

> Land clearing: eliminating existing vegetation, clearing and levelling the land so that it can be prepared for planting;

> Soil preparation - involves operations that depend on the type of soil, previous use and compaction, including ploughing, harrowing, subsoiling, chemical weeding, etc;

> Planting: is carried out at varying spacings, with 6 to 9 m^2 of land per tree; it includes operations such as alignment (when subsoiling is not carried out), digging and fertilising, combating ants, applying pre-

[7] Note: AIB is an artificial rooting-inducing hormone (a type of synthetic auxin).

emergent herbicide, transporting and distributing seedlings, planting and replanting seedlings, irrigation, combating pests and diseases; completion of the guarantee period for the seedlings to take hold and survive, which varies from company to company and can be considered to be just the first month after planting or up to the entire first year; after that, the maintenance period begins;

> Maintenance: generally considered to be the first two years after planting; during this period operations such as ant control, fertilisation, combating pests and diseases and controlling competition through mowing and mechanical or chemical weeding are carried out;

> Costing: this is the period of least activity and when the risks of fire are greatest; the main operations at this stage are the control of ants, other pests and diseases, and sometimes weeding and fertilising;

> Harvesting: a period that usually begins around six months before the actual felling and includes pre-harvest ant control operations, pre-felling mowing, road repairs, tree felling, delimbing and processing, hauling the wood to the roadside, loading and transporting the wood, which can be directly to the mill, or include a depot for changing transport vehicles, which is necessary when the vehicles travelling through the forest have to be of a different type to those used for the main long-distance road transport.

3.2.3 Silvicultural activities and operations in phase 2ª - Reforestation:

> Regrowth management: involves operations to ensure the survival and good growth of the regrowth of the vines and the elimination of surplus shoots, such as: combating ants, chemical weeding, fertilisation, elimination of excess shoots (usually 1 to 3 shoots are left, depending on the gaps around the vine);

> Maintenance: same as deployment;

> Costing: same as implementation;

> Harvest: same as implantation.

3.2.4 Silvicultural activities and operations in phase 3 - Reform:

> Seedling production: same as implantation.

> Clearing the land: this activity is generally restricted to eliminating sprouting and sometimes lowering stumps and eliminating excess harvest residues (branches and tree tips); exceptionally there is a need for other operations, which happens when there is a lot of residue from the previous harvest, or the occurrence of large stumps (in the case of old-growth forests with more than one sprouting phase); land clearing for reforestation is almost exclusively about eliminating sprouting, sometimes includes eliminating excess harvest residues and lowering stumps, and rarely includes clearing and levelling the land to prepare it for planting; various operations have been used to eliminate harvest residues in land clearing for reforestation, such as lowering stumps, manual or mechanical removal, ploughing with extra-heavy disc harrows that crush the residues, or with a knife roller, or controlled burning of the residues, which is currently banned due to the environmental impact it causes;

> Soil preparation - involves operations that depend on the type of soil and compaction, usually restricted to chemical weeding and subsoiling, which is considered minimum forest cultivation, but there are variations with ploughing and harrowing between the stump lines;

> Planting: includes operations such as alignment (when furrowing is not carried out), hoeing and fertilising, combating ants, applying pre-emergent herbicide (little used due to the reduced effect of harvest residues in minimum cultivation), transporting and distributing seedlings, planting and replanting seedlings, irrigation, combating pests and diseases;

> Maintenance: same as deployment;

> Costing: same as implementation;

> Harvest: same as implantation.

At the same time, in all phases, general infrastructure maintenance operations are carried out, such as the upkeep of roads, works of art, buildings, fences, etc.; as well as fighting fires, competition, pests and diseases, which are carried out when indicated by monitoring.

Seed production areas are treated in a similar way to regular high-stem areas, except that the aim is not to harvest timber, but to produce the maximum amount of seeds.

Clonal gardens are treated and managed similarly to vegetable gardens, with irrigation, fertilisation and pest and disease control; they are areas where clones are planted with minimum spacing and kept at the height of a vegetable, receiving intensive treatment to produce shoots, which are used to produce clonal seedlings; there are already experimental hydroponic clonal orchards and *in vitro* reproduction is increasingly being considered for commercial use, although there are still operational problems *in* producing shoots on a large scale.

3.3 Eucalyptus cultivation costs

The operating costs indicated by Bahia Sul (II RITEC, 1993), were US$ 630/ha for renovation and cultivation up to 6 months of age, and with minimum cultivation starting soon afterwards, the costs fell considerably; for harvesting (cutting + baling), the costs were US$ 4.17/st (about US$ 6. 7/m(3)), for transport they were US$ 1.9/st (about US$ 3. 7/m(3)).7/m^3), while transport costs were US$ 1.9/st (around US$ 3.04/m^3), including road repairs, road maintenance and loading the wood, considering an average distance from the mill of 49 km; harvesting costs have changed with the mechanisation of operations, but are still close to this.

The costs of establishing a eucalyptus plantation in the southeast of Goiás, obtained by Furtado et al (2022), came to R$ 6,381.71, equivalent to approximately US$ 1,276 per hectare planted (Table 6).

TABLE 6 - Costs of planting *E. urograndis* in the south-eastern region of Goiás

Activity	Cost components	R$/ha
Cleaning the area	Sprayer	45,00
	Herbicide Glyphosate 480 (Roundup Transorb)	200,00
	Labour cleaning the area	37,50
Soil preparation	Soil analysis	45,00
	Road upgrading	540,00
	Subsoiling + Fertilising	270,00
Soil correction	Limestone	375,00
	Freight of limestone to the farm	600,00
	Labour distribution of limestone	75,00
	Dispensing machine	90,00
	R$50.40 NPK 20-05-20	375,00
Fighting ants	Formicide Bait	14,00
	Ant Fighting Labour	75,00
Planting	Purchase of seedlings (planting and replanting 5%)	874,65
	Planting labour	150,00
	Cart - kite	720,00
Cultivation	Replanting labour	75,00
Activity	**Cost components**	**R$/ha**
Chemical weeding (Goal Herbicide) + Sprayer		90,00
Carpina chemical labour		75,00
Service Technical Assistance		800,00
EFFECTIVE OPERATING COST (EOC)		5201,55
Other expenses (10% of the COE for any expenses not included)		520,16
Opportunity cost of land		660,00
TOTAL OPERATING COST (COT)		6381,71

Source: Furtado et al (2022)

Moreira et al (2022) obtained the results presented in the income and costs of a project to plant *Eucalyptus urophylla,* with planting spacing of 3 m x 2.5 m, with two 6-year rotations, for the production of firewood, in the region of Rio Verde, GO, with and without resources from the ABC Programme. The production considered was 310.10 m^3 per ha and 262.00 m^3/ha in the first and second rotation, respectively.

TABLE 7 - Costs and revenues of the *Eucalyptus urophylla* project, without and with ABC Programme resources in 2018

Distribution of remuneration (R$/ha)	ABC resources Without	With
Gross production revenue	64.472,25	64.472,25
Gross financing income		05.337,22
Total cost of transport	16.762,79	16.762,79
Total harvest cost	15.866,27	15.866,27
Total cost of forestry	10.149,06	10.149,06
Total cost of financing		05.929,45
Total cost of production tax	3.372,54	3.372,54
Total IRPF cost	3.128,68	3.128,68
Total rental cost	7.200,00	7.200,00
Total cost of interest on equity	6.050,64	4.555,84
Net profit = NPV	1.942,28	2.844,84

Source: Moreira et al (2022).

The Campo Futuro Project (CNA, 2022) carried out a survey of large-scale eucalyptus timber production costs in Ortigueira (PR). The production costs were established on the basis of a 125-hectare modal property, the most common in the region, with an Average Annual Increment (Ima) of 35 m3/ha/year, two thinnings at the 6th and 10th years and harvesting at the 15th. Eucalyptus plantation costs, involving pre-planting, planting and post-planting operations, totalled R$9,471/ha.

The evolution of m^3 prices for Eucalyptus logs between 2014 and 2021 is shown in Figure 15, reflected in the cost of purchasing wood for the industry, which exceeded US$ 90 per m3 for lamination in 2021, and for pulp, the value was over US$ 85 per m3 in the same year.

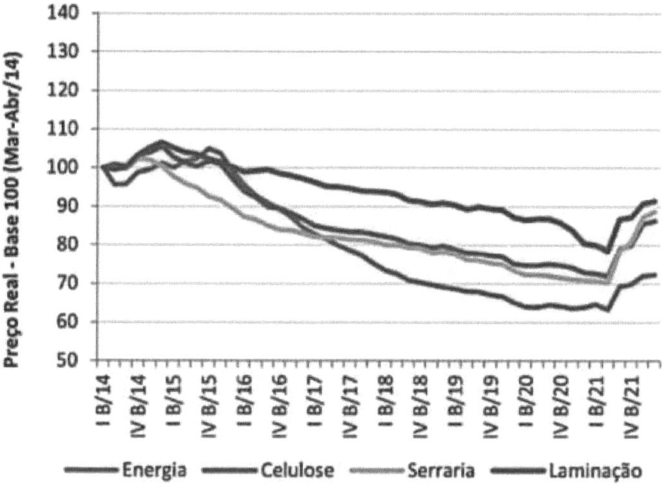

Figure 15 - Price of standing eucalyptus wood in Brazil - log assortments: energy: < 8 cm; pulp: 8-15 cm; sawmill: 16-25 cm; lamination: 25-35 cm; and special lamination: > 35 cm. Source: b.Forest / STCP (2022).

3.4 Additional comments on *Eucalyptus*

3.4.1 Planting spacing

Planting spacing varies greatly according to the purpose of the wood produced and the cultivation system adopted; the list below shows some examples of spacing used in eucalyptus plantations:

> Spacing of 3 m between rows X 2 m between plants in the row - used to produce small-sized wood for energy, particles or fibres in regions with

moderate growth in a coppice system and rotation of 7 to 8 years;

> Spacing of 3 m between rows and 3 m between plants - used to produce small-sized wood for energy, particles or fibres in regions where growth is vigorous and rotation lasts 5 to 7 years;

> Spacing of 5 m between rows X 2.7 m to 3 m between plants in the row - used to produce large timber for sawmills and laminations, with pre-commercial thinning at 3 years, 1st commercial thinning at 7 years, 2nd commercial thinning at 11 years and clear-cutting at 14 years, and the areas can be used for agricultural or livestock production in the first 18 months (Agroforestry or Silvipastoral Systems, respectively).

> Strip planting (renques) of 3 rows of eucalyptus 3 metres apart and 2 to 3 metres between plants in the row, with 14 to 20 metres between the strips - used to produce small timber in a silvipastoral or agrosilvicultural system, with rotations of 7 years, but can be managed in longer rotations to produce larger timber.

3.4.2 Multiple use

Eucalyptus can also be used for windbreaks, ornamentation, watershed protection, the recovery of degraded areas, bee pastures and more.

The multiple use of forests is increasingly being encouraged by pressure from the environmental field, through NGOs and some sectors of society. SOARES (2002-c), studying the multiple use of wood, concluded that the utilisation of wood for multiple purposes had greater economic results in the situation studied. Examples of uses associated with eucalyptus cultivation are: planting annual crops between the eucalyptus rows, during the first two years with species such as corn and soya; beekeeping, due to the abundant flowering enjoyed by bees; and livestock farming, which is being avoided by many companies, as there is a claim that grazing in eucalyptus forests causes excessive soil compaction, which is detrimental to its main objective, the production of wood.

3.4.3 Physicomechanical characteristics of wood

According to CARVALHO (2000), basic density is a good indicator of industrial yield in pulp due to its high correlation with yield in *eucalyptus* pulp production; the density some *Eucalyptus* woods are: *E. dunnii* 0.493 g/cm^3, E. urophylla 0.564 g/cm3, *E. grandis* 0.401 g/cm3 and *E. saligna* 0.450 g/cm3; hybrids of *E. grandis* and *E. urophylla* 0.493 to 0.515 g/cm3; species that are close to the optimum range for short fibre pulp production, which is around 0.450 to 0.550 g/cm3. The physicomechanical characteristics of *Eucalyptus* vary from species to species, some of which are shown in Table 8:

TABLE 8 - Physical and mechanical characteristics of *Eucalyptus* wood

CHARACTERISTICS	*E. urophylla*	*E. saligna*
Apparent specific mass (g/cm2)	0,83	0,69
Species Volumetric contraction (%)	20,4	23,4
Maximum static bending strength (MPa)	94,9	77,4

Source: CARVALHO (2000).

3.4.4 chemical composition of wood and fibre dimensions

The cell wall of wood cells is made up of three main components: cellulose, hemicellulose and lignin. Wood of the *Eucalyptus* genus is basically made up of 50% cellulose, 25% hemicellulose and 25% lignin, according to BLAND *apud* SCARPINELLA (2002); according to the latter, the elemental composition of dry wood on average is 50% carbon, 44% oxygen, 6% hydrogen, 0.1% nitrogen and 0.3% ash.

CARVALHO (2000) found results for the chemical composition of the wood that were very close to the values found in previous studies with the parental species for hybrids *of E. grandis* and *E. urophylla. urophylla* hybrids, with total extractives of 3.65 to 3.69%, lignin content of 22.20 to 22.59%, holocellulose content of 73.76 to 74.10%; in an analysis of the fibres, he found average values for their dimensions of: length 1.075 mm; width 17.31 µm; lumen diameter 8.625 µm; wall thickness 4.34 µm; the bleached pulp yields found by the author were over 50%.

JERÔNIMO *et al.* (2000), studying the cooking of *Eucalyptus saligna*, obtained an average yield of 52.2% in purified bleached pulp.

3.4.5 Brazilian research on *Eucalyptus*

There are currently 76 forestry engineering colleges in Brazil, the oldest of which run postgraduate courses, such as: Curitiba (UFPR) - up to post-doctorate; Viçosa (UFMG) - up to doctorate; Santa Maria (UFSM) - up to doctorate; Piracicaba (ESALQ) - up to doctorate; Rio de Janeiro (UFRRJ) - up to doctorate. Most of the schools have forestry research organisations linked to them, such as FUPEF (PR), IPEF (SP) and SIF (MG).

Embrapa has a unit exclusively dedicated to forestry research in Colombo, PR, which has carried out research with eucalyptus in Paraná; it carried out some of the first research with *E. dunnii* in Brazilian territory.

Several states have forestry research organisations or units that carry out studies on eucalyptus, such as São Paulo, Rio Grande do Sul and Paraná.

Practically all the large Brazilian companies that work with eucalyptus, especially those with more than 50,000 hectares under cultivation, have well-equipped research centres with highly qualified staff.

Brazil has invested heavily in eucalyptus research, mainly in genetics, with the aim of increasing productivity and reducing operating costs. Through the GENOLYPTUS PROJECT, it has formed the largest and most complex forestry experimentation network in the world, stretching from the Amazon to Rio Grande do Sul: the Brazilian *Eucalyptus* Genome Research Network, officially launched on 20 February 2002 in Brasilia, according to an article in REVISTA DA MADEIRA (May 2002), which reports:

> The Brazilian Eucalyptus Genome Research Network brings together scientists, laboratories and companies from all over Brazil to research the best ways to increase production, reduce industrial pollution and create eucalyptus trees that are more resistant and suitable for different uses.
>
> There are 12 companies, seven universities and Embrapa working on the project, made possible with resources from the Yellow Green Fund of the Ministry of Science and

Technology. In the first phase, planned for five years, the Genolyptus Project will consume around R$12 million reais. Companies will contribute 30 per cent of this total and the government the rest, through the work of scientists from public universities and scholarship holders, laboratory facilities and resources from the Yellow Green Fund. At the end of 2001, the Fund released R$2 million to start the work and has already secured a further R$2 million for this year. The researchers estimate that the first practical results will appear after the first eucalyptus seedlings have grown, in around two years. According to the coordinator of Genolyptus, Dario Graltapaglia, who is a professor and works at the Genomic Biotechnology Laboratory at the Catholic University of Brasilia, "the aim, after this initial five-year phase, is to make it a permanent project, where applied research is a constant for the companies involved, always seeking better results".

Aracruz has achieved yields of up to 18 tonnes of pulp per hectare per year with *E. urophylla* and *E. grandis* hybrids, according to CARVALHO (2000), whereas at the start of the company's activities it was no more than 6 tonnes. These figures, if extended to all Brazilian conditions, could lower production costs considerably. The GENOLYPTUS project may not provide this, as it also depends on soil, climate and silvicultural conditions, but Brazilian competitiveness, through cooperation between companies, universities and research centres, could make pulp production unviable in many parts of the world. On the other hand, there is a risk that high productivity will be unsustainable in the long term if nutrients are not replenished, especially from sustainable natural sources.

The main method of eucalyptus improvement is recurrent selection with seeds harvested from wild populations to produce seedlings and then form base populations, from which the best individuals are selected for progeny testing, where the best are selected and reproduced by cloning. Many progenies turned out to be hybrids, with very high productivity, especially E. grandis with E. urophylla; artificially controlled hybridisation was then carried out, obtaining new strains that are also cloned. The use of transgenics has also been included in the last decade, but there is great pressure against this technology from both society and the market.

3.5 Regional Physiognomies

Eucalyptus forests for pulp production were concentrated in just a few regions of the country until the beginning of this century, such as the state of São Paulo and the north of Espírito Santo, as can be seen in Table 1, but they have become widespread in other regions.

TABLE 1 - LOCATION OF EUCALYPTUS FORESTS BY BIOME

BIOME: / COMPANY:	AMAZON	ATLANTIC FOREST	EcossiT. cosTEiRos	cERRADo	SOUTHERN FIELDS
Aracruz (Suzano) Bahia Sul (Suzano) Celmar Cenibra IP Jari Klabin Riocel (CMPC) Ripasa (Suzano) Suzano VCP (Suzano) Veracel					

Source: Localisations made using information on the companies' websites - not all of the biome domains in which the companies operate may have been identified.

With the exception of the Mato Grosso Pantanal and the Northeastern Caatinga, eucalyptus forests planted for pulp production are located in regions where all the other Brazilian biomes occur: the Amazon, Cerrado, Atlantic Forest, Southern Fields and Coastal Ecosystems.

The largest area planted with eucalyptus is in the Southeast (46%), followed by the South and Centre West (18% each) (Figure 16).

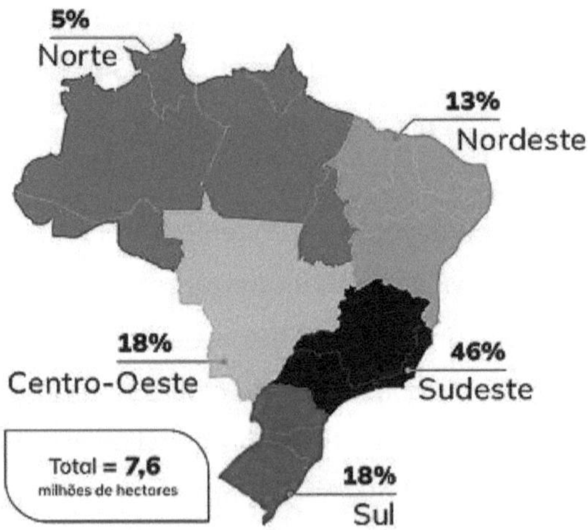

Figure 16 - Eucalyptus planted area by region (%), 2022.

The locations where pulp and paper industries that use eucalyptus wood are concentrated in Brazil are identified on the map in Figure 17 and the history of plantations by state is shown in .

***GEOGRAPHIC DISTRIBUTION OF PULP* AND PAPER PRODUCERS**

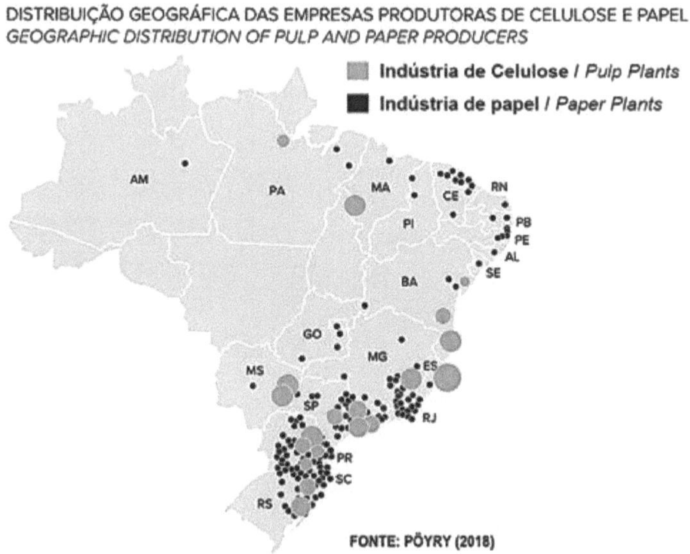

Pulp* Industry / *Pulp Plants

Paper* industry / *Paper Plants
SOURCE: PÖYRY (2018)
Figure 17 - Location of pulp mills. Source: IBÁ (2019).

In 2012 there were 5.3 million hectares of eucalyptus plantations in the country and until 2017 there was little increase, but from then on there was a considerable increase of more than 1 million hectares until the following year and of more than 800,000 hectares for 2019, rising to 7.5 million hectares, which has remained almost unchanged since then (Table 9).

TABLE 9 - Changes in the area of eucalyptus planted by state

State	2012	2014	2016	2018	2020	2022
Minas Gerais	1.438.971	1.400.232	1.390.032	1.970.063	2.233.762	2.201.801
Mato Grosso do Sul	587.310	803.699	877.795	1.093.805	1.022.521	1.110.484
São Paulo	1.041.695	976.186	946.124	1.035.874	945.114	999.987
Bahia	605.464	630.808	612.199	585.258	637.765	650.332
Rio Grande do Sul	284.701	309.125	308.178	426.371	581.338	589.674
Paraná	197.835	224.089	294.050	255.955	448.818	442.222
State	**2012**	**2014**	**2016**	**2018**	**2020**	**2022**
Santa Catarina	106.588	112.944	116.240	219.199	307.229	317.724
Maranhão	173.324	211.334	221.859	200.612	268.912	290.846
Holy Spirit	203.349	228.781	233.760	225.520	260.170	258.533
Pará	159.657	125.110	133.996	151.888	167.354	166.215
Goiás	115.567	124.297	127.201	127.201	161.940	154.380
Mato Grosso	184.628	187.090	185.219	187.947	120.489	130.003
Tocantins	109.000	115.564	116.798	149.886	98.988	100.477
Amapá	49.506	60.025	65.026	67.826	68.462	362

Others	46.568	49.369	45.307	87.983	84.395	172.060
Total	**5.304.163**	**5.558.653**	**5.673.784**	**6.785.388**	**7.407.257**	**7.585.103**

Source: IBÁ (2023).

Brazil's biomes are distributed as shown in Figure 18 and are described below.

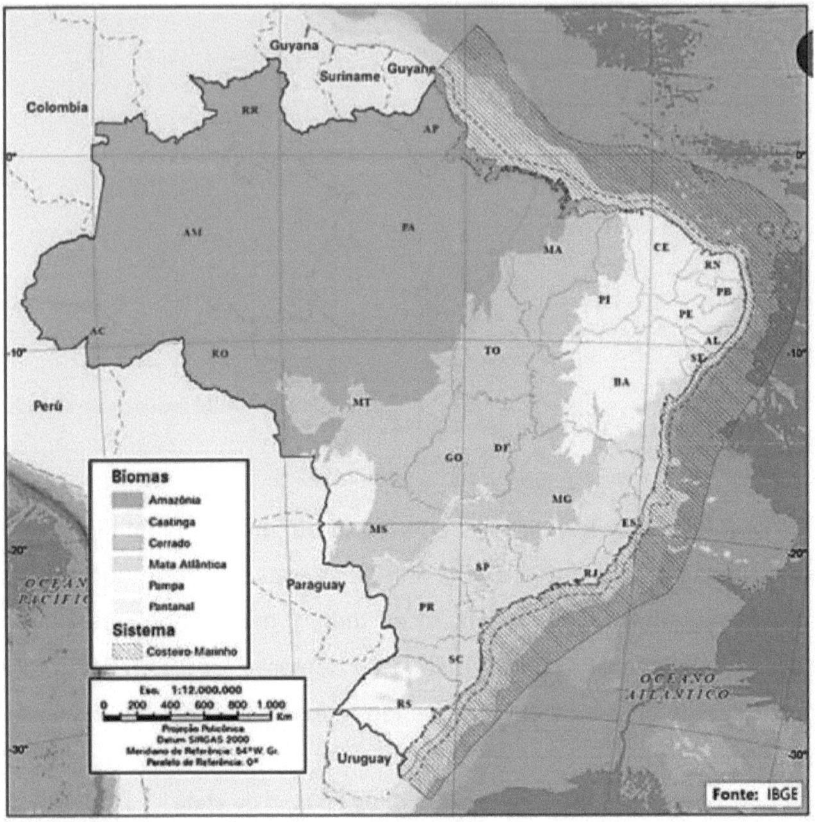

Figure 18 - Map of Brazilian biomes. Source: IBGE.

3.5.1 Amazonia

The Amazon covers 40 per cent of South America. It is considered to be the largest tropical rainforest in the world, whose domains occupy around 3.7 million kilometres of Brazilian territory[2], covering the states of Acre, Amapá, Amazonas, Pará, Rondônia, Roraima and part of the states of Maranhão, Tocantins and Mato Grosso, equivalent to 1/3 of the world's tropical rainforest reserves. In addition, it holds 20% of the world's available fresh water and an unmeasured mineral heritage.

It competes with the Atlantic Forest in terms of both ecosystem diversity and the number of species. It is currently considered the largest genetic bank on the planet.

The geological and relief diversity has resulted in the formation of a variety of soil classes, under a warm equatorial climate that ranges from humid to super-humid. However, most of the soils developed on sedimentary substrates, resulting in low natural fertility. The exuberance of the Amazonian ombrophilous forest is not a consequence of the fertility of the soils, which is low; in fact, it is due to ecological adaptation and the warm, rainy climate.

Most Amazonian ecosystems are closed-cycle; in other words, everything the forest produces, it consumes itself. The forest's stock of carbon and other elements and substances is practically constant.

The forest acts as a storehouse of carbon and water and thus contributes to the global climate balance. Because they are closed ecosystems in equilibrium, called climax, on poor soils, they are extremely sensitive to exploitation and the removal of anything. Amazonian ecosystems include: dryland forests, flooded forests, floodplains, igapós, open fields and savannahs.

Jari has its plantations on the border between the states of Amapá and Pará, in a region of terra firme forests with soils that are highly susceptible to erosion. The main feature of the local topography is the existence of small hills with flat plateaus. The company plants on the plateaus, avoiding any activity on the slopes. The access roads to the plantation areas require special drainage works to prevent erosion, such as gullies on steps and slopes with humps from space to space to break up the speed of rainwater.

3.5.2 Southern fields

The southern grasslands are generically called "pampa", an indigenous term for "flat region". This name, however, corresponds to only one type of grassland, most commonly found in the centre and south-west of the state of Rio Grande do Sul, reaching into Uruguay and Argentina. Other types known as campos de cima da serra are found in transition areas with the *Araucaria angustifolia* domain and the southeastern campos da serra are found bordering the coastal ecosystems and the pampa. In other areas there are also savannah-like grasslands.

Through empirical observation of the evolution of the landscape and deduction from bibliographic reports, mainly by LANI *et al.* (1999), it is assumed that the natural pampas of Rio Grande do Sul practically no longer exist and seem to have been formations of dirty grasslands and savannas with edaphic origins, but this needs scientific proof. With anthropogenic action, associated with low soil fertility, grazing pressure and the practice of clearing fields with fire, the shrub vegetation has been eliminated, with taller vegetation remaining only on the edges of watercourses.

The soils are sandy and highly erodible in areas with a slope of 2% or more. On the Campanha plateau, in the centre and southwest of Rio Grande do Sul, there are dark red podzolic soils, where there is the phenomenon of sandification due to excessive and predatory use of the soil.

The grassland vegetation is uniform on the coxilhas and is lower and sparser on the tops, becoming denser and richer on the slopes, with a predominance of grasses, composites and legumes. Along the watercourses there is a riparian forest with numerous tree species of commercial interest and marshes form in the flatter areas, especially at the springs.

Riocel concentrated its activities in the region of the fields between Lagoa dos Patos and the Jacuí River, in the east of the state, between irrigated rice plantations and pastures. Its successor, CMPC, expanded its plantations and acquired areas planted by other companies in the southern half of Rio Grande do Sul, with more than 2 million tonnes of pulp produced by 2022. The soils there are sandy loam to sandy loam with low fertility and high erodibility. The topography varies from flat to slightly undulating. The region where the company operates has some transition points to the Atlantic Forest.

3.5.3 Cerrado

The Cerrado biome is a tropical savannah-type ecosystem of edaphoclimatic origin, with an annual dry season and a rainy season and soils low in micronutrients and high in aluminium. The Cerrados are distributed across the central Brazilian plateau, in the states of Goiás, Tocantins, Mato Grosso, Mato Grosso do Sul, part of the states of Minas Gerais, São Paulo, Bahia and the Federal District, covering almost 2 million km^2. There are also some areas of cerrado in the ecotones between the Amazon, Atlantic Forest and Caatinga biomes.

There are several physiognomies: cerradão, typical cerrado, campo cerrado, campo sujo de cerrado and campo limpo, which have height and plant biomass in decreasing order, with cerradão being a type of forest formation, while the other physiognomies are savannas.

The typical cerrado has low, sparse trees, up to twenty metres high, among shrubs, sub-shrubs and undergrowth with a predominance of grasses, making up two strata: upper - made up of trees and shrubs with deep roots; lower - made up of herbaceous plants, dominated by creeping grasses with shallow roots. The light intensity reaching the lower stratum is high. The lower layer dries out during the dry season, favouring the spread of fires. The characteristic vegetation has crooked trunks of low size, twisted branches, thick bark and thick leaves. The Brazilian Cerrado is considered to be the savannah with the greatest biodiversity, with diverse ecosystems.

Until the 1950s, the Cerrados remained almost unchanged. From the 1960s onwards, with the internalisation of the capital and the construction of the road network, livestock farming and extensive agriculture advanced over them with soya, rice and wheat crops.

During the 1970s and 1980s, the agricultural frontier advanced, resulting in 67 per cent of the Cerrado being occupied, leaving only 20 per cent of the original area conserved.

Companies that planted eucalyptus in regions where savannahs occur were Ripasa, Suzano and VCP, now all owned by Suzano, as well as Eldorado and Arauco.

3.5.4 Coastal Ecosystems

There is a rich mosaic of ecosystems on the Brazilian coast, of high environmental relevance, made up of seas, estuaries, islands, mangroves, sandbanks, dunes, beaches, cliffs, rocky coasts and coral reefs, due to climatic and geological differences.

The vegetation is extremely diverse.

Mangroves fulfil essential functions in the reproduction of marine life.

The Amazon coast runs from the mouth of the Oiapoque River to the delta of the Parnaíba River. It has a large expanse of mangroves and tidal floodplain forests, dune fields and beaches. It has a rich diversity of crustacean, fish and bird species.

The northeastern coast stretches from the mouth of the Parnaíba River to the Recôncavo Baiano. It is marked by limestone and sandstone reefs, shifting dunes and others covered in vegetation. There are mangroves, sandbanks and forests. The waters off the northeastern coast are home to endangered manatees and turtles.

The south-east coast runs from the Recôncavo Baiano to São Paulo. Parts of the area are among the most densely populated and industrialised in the country, while others have a low population density, such as the south of Bahia. Characteristic landscapes are cliffs, reefs and monazite sand beaches. It is dominated by the Serra do Mar mountain range and the coast is indented by bays and inlets. The most important ecosystem is the restinga forest. This part of the coast is home to the endangered golden sloth and the golden lion tamarin.

The south coast begins on the border between São Paulo and Paraná and ends at the Chuí Stream in Rio Grande do Sul. With many marshes and mangroves, the region's ecosystems are rich in birds and other species such as the endangered marbled rat and otters.

Due to the high anthropogenic impact that already exists in the coastal region, the Ministry of the Environment, in cooperation with the Interministerial Council for the Sea, state governments, IBAMA and other institutions, is trying to organise and protect ecosystems with the implementation of the National Coastal Management Plan (PNGC).

Companies operating in the region are Aracruz and Bahia Sul, in the flat, sandy areas of northern Espírito Santo and southern Bahia, now both owned by Suzano.

3.5.5 Atlantic Rainforest

It is the fifth most endangered area in the world and rich in endemic species. It is made up of a diverse mosaic of ecosystems, with different structures and floristic compositions depending on the differences in soil, relief and climate that exist throughout its area of occurrence, from Rio Grande do Sul to Rio Grande do Norte and from sea level to over 1500 metres above sea level.

Its current area is highly fragmented. Most of the remnants of the original vegetation cover are located in areas that are difficult to access. The preservation of these remnants has ensured the containment of slopes along the Atlantic side of the Serra do Mar. Around 7.3 per cent of the original Atlantic Forest area remains, which has some of the most exuberant landscapes in the country. The Atlantic Forest is home to water sources that supply around 70 per cent of Brazil's population. The legal protection of the existing ecosystems is very high, and the suppression of natural vegetation is prohibited in almost all of them.

The main forest ecosystems and associated ecosystems of the Atlantic Forest are: dense ombrophilous forest from Rio Grande do Sul to Rio Grande do Norte; deciduous and semi-deciduous seasonal forests in Rio Grande do Sul, Santa Catarina, Paraná, São Paulo, Rio de Janeiro and Espírito Santo; semi-deciduous seasonal forests of Mato Grosso do Sul, occurring in the valleys of the tributaries of the Paraná River, the valleys of the Parnaíba and Grande rivers and their tributaries in Minas Gerais, the valleys of the Paraíba do Sul and Jequitinhonha rivers, intermediate rivers and their tributaries in Minas Gerais and Bahia and the limited coastal regions of the Northeast, contiguous to the ombrophilous forests; the mixed ombrophilous forest in the states of Rio Grande do Sul, Santa Catarina and Paraná; and the araucaria enclaves in the states of São Paulo, Rio de Janeiro and Minas Gerais.

The floristic formations associated with the biome are: mangroves, restinga vegetation and coastal island vegetation; cerrado enclaves, fields and highland fields within the above areas; hilltop and hillside forests in the Northeast (brejos and chás), particularly those in the state of Ceará, with emphasis on the Ibiapaba and Baturité mountains and the Chapada do Araripe; and the native vegetation formations on the island of Fernando de Noronha.

Of the large companies in the pulp and paper sector, with the exception of Jarí and Celmar, all the others have production forests in areas of the Atlantic Forest biome, in several of its ecosystems, starting with VCP, Aracruz and Bahia Sul in Bahia, the latter two also in Espírito Santo, in regions generally with little slope and sandy soils; Veracel in the south of Bahia; Cenibra in Minas Gerais, in a mountainous and more clayey region; VCP in São Paulo, where Suzano and Ripasa are also located, in diverse soils and topographies; further south are Klabin in the Araucaria region and Riocel in Rio Grande do Sul (now CMPC). Environmental conditions vary enormously in these areas, both in terms of soil, from moderately fertile to poorly fertile, from sandy to clayey; and in terms of climate and topography, from flat to mountainous; and seasonal to super-humid climate, from almost equatorial to practically temperate.

CHAPTER 4

4 Impacts of eucalyptus cultivation

The effects of eucalyptus plantations have been discussed all over the world and are the subject of books such as "Impacto Ambiental do Eucalipto" (Environmental Impact of Eucalyptus) by Walter de Paula Lima (1987), which cites symposiums and various situations in which the subject has been addressed.

To summarise, the problems pointed out in relation to the production of eucalyptus wood for the Brazilian pulp and paper industry are as follows:

- Soil depletion due to excessive nutrient consumption;
- Drying up of water sources due to the high demand for water;
- Allelopathy and the consequent reduction in the diversity of flora, due to the release of chemical substances that are toxic to other plants and soil micro-organisms, further reducing fertility;
- Reduction in the diversity of fauna, due to the reduction in the diversity of flora and the fact that it is exotic and not to the taste of Brazilian fauna;
- Excessive use of agrochemicals with consequent soil and water contamination;
- Land concentration.

Land concentration is a reality, given the huge rural properties owned by companies, which are necessary for the plantations that make them self-sufficient in raw materials to comply with current legislation (IBAMA Normative Instruction 001/80). One of the ways to avoid more serious problems is to limit the percentage per municipality that can be dedicated to the activity, an action that Espírito Santo has already been forced to adopt due to the large number of medium and large forestry companies operating in the state.

It's true that eucalyptus consumes nutrients from the soil, like any other crop, but it's neither the biggest consumer nor the smallest, as you'll see below. It's also true that it consumes a lot of water and reduces biodiversity, as other monocultures do. The allelopathy of eucalyptus has not yet been fully studied, but it seems to be more a consequence of the high level of competition it has with other plants than the effect of allelopathic substances (LIMA, 1993). Agrochemicals such as insecticides and fertilisers are used in eucalyptus farming, as in other crops, and need to be controlled, but there are advantages to forestry over annual crops, as trees generally have greater resistance to pests and diseases than herbaceous plants, and eucalyptus has been shown to be less demanding in terms of nutrients.

The important thing is to quantify these effects and take steps to minimise or eliminate them, because the benefits of eucalyptus plantations also exist and, as you can see from the previous chapters, they are quite significant, and it is difficult for the Brazilian forest-based industry to do without eucalyptus species today, given the technological advances in eucalyptus cultivation compared to the silviculture of other species that could be used for pulp production; It is even more difficult for Brazilian society to give up the trade surplus, jobs, salaries, social and cultural benefits that the pulp and paper industry brings to the country.

Among the benefits of eucalyptus cultivation, cited by REVISTA DA MADEIRA (September 2001), are:

- Reduces pressure on native forests;
- Absorbs atmospheric CO_2;
- It contributes to regulating the flow and quality of water resources;
- It can be used as a windbreak;
- It supplies raw materials to the pulp and paper, board, chipboard, sawn timber and fibre industries, among others;
- It generates direct and indirect jobs;
- Maintains soil cover through litter, reducing the risk of erosion;
- Contributes to nutrient cycling.

Some scientific studies on the subject are covered in the following sections, including the impacts of plantations, roads and infrastructure works. The social, cultural and economic impacts are covered in the Introduction and in Chapter 1, where the Brazilian pulp and paper sector is characterised.

4.1 Nutrient intake

Eucalyptus shows variations in chemical composition between species and between locations where it is planted; in addition, due to variations in climate, soil and nutrients precipitated from the air by rainfall, the availability of nutrients varies enormously, which merits specific studies into the replacement needs for each species used in each region. Some data from studies carried out are shown in Table 10 to Table 13, including other plant genera for comparison. What can be observed is that the amounts of nutrients removed from the soil by *Eucalyptus* harvests, per tonne of dry matter harvested, are much lower than those removed by agricultural species such as soya, maize, coffee and slightly higher than sugar cane.

TABLE 10 - Chemical characteristics of some forest soils in the depth from 0 to 20 cm

Soil	Locai	pH		M.O.	P	K		Ca	Mg	S	CTC	V	m
		H,0	CaCl,	$g.dm^3$	$mg.dm'^3$			- mmol .dm E !				...	% -
AQ[1]	Itatinga-SP	-	3,7	12	9	0,5	3		1,5	5,0	37	13	-
AQ[1]	Lençóis Paulista-SP	-	3,8	14	4	0,4	4		1,0	5,4	34	15	-
AQ[1]	Altinópolis-SP	-	4,0	20	4	0,3		1,50,3		2,1	56	3	84
LR[2]	Capào Bonito-SP		3,8	38	4	0,3	2		1,0	3,3	101	3	-
LVE[J]	Bocaióva-MG	4,8		21	2	1,2	4	2,0		7,2	22	-	
LVA[4]	Itatinga-SP		3,4	19	5	0,6	3	2,0		5,6	70	9	-
LVA[4]	Itamarandiba-MG	4,2	-	33	1	0,6	0,05		0,01	0,7		-	-
LA[5]	Laranjal do Jari-AP	-	3,6	12	3	0,4		2	2,0	4,4	47	Il	68
PVA[6]	Teixeira de Freitas-BA	5,5		*	0,6	0,8		0,60,6		2,0	36		
Average		4,8	3,7	19,5	4	0,5	2		1,0	4,4	42	10	76

(1) Quartz sand; (2) purple latosol; (3) dark red latosol; (4) yellow red latosol; (5) red latosol; (6) yellow red podzolic.
Source: SILVEIRA *et al.* (2001).

According to SILVEIRA *et al.* (1995c) *apud* SILVEIRA *et al.* (2001), the distribution of the main soil types used in eucalyptus plantations is as follows: 64% dystrophic or alkalic latosols; 16% dystrophic or alkalic podzols; 10% cambisols and lithosols; 5% quartz sand; 2.5% purple earth, podzols and eutrophic latosols; and 2.5 others. It can be seen in

Table 10 that the soils listed are poor in phosphorus and around 50% are poor in potassium, with the latosols being slightly more fertile than the others, but eucalyptus productivity is only average in these locations considering the commercial plantations, however, the low-fertility podzolic soils of Teixeira de Freitas in Bahia support the high productivity levels of the Aracruz and Bahia Sul eucalyptus plantations, indicating that soil fertility is not a limiting factor for the crop.

Table 11 shows that although agricultural harvests remove a small amount of biomass, their nutrient concentration is 5 to 20 times higher than that of wood. Depending on the type of crop, 7 or 14 agricultural harvests are carried out during a eucalyptus cycle, resulting in nutrient exports from agricultural harvests that are many times higher than those from timber harvests.

The leaves concentrate the most N, P, K, S and Mg; the bark is rich in P, Ca and Mg; the branches have high concentrations of N and K; while the wood has the lowest concentrations of all the nutrients, but the highest in absolute terms.

The high concentration of nutrients in leaves, bark and branches means that, whenever possible, they should be returned to the soil, unlike what was done until the mid-1990s, when part of the bark was burnt to produce energy and the rest was dumped in waste areas, while branches and leaves were burnt in the field to facilitate new plantings, volatilising part of the nutrients and making another part leachable before the plants could use them.

TABLE 11 - Concentration of macronutrients in plant parts removed by harvesting in Kg / Ton / cycle

SPECIES	AGE (YEARS)	COMP(*)	BIOMASS (T/HA)	AMOUNT OF NUTRIENTS (KG/TON) N	P	K	CA	MG
E. grandis	10	M	160	0,8	0,02	0,6	1,7	0,5
		C	23	3,0	0,27	3,0	3,1	1,0
E. grandis	6	M	62	2,3	0,11	0,8	0,5	0,2
		C	11	3,4	0,76	6,0	7,9	1,4

C.citriodora	9		65	1,8	0,56	1,4	1,1	0,6
		M + C M	17	3,1	0,42	4,6	8,5	3,2
E. grandis	7	C	227	1,5	0,13	1,1	2,7	0,5
E. saligna	10	M	129	0,9	0,23	0,6	0,8	0,1
		C	8	3,1	1,51	6,0	9,5	3,9
P. oocarpa	8	M	70	1,3	0,12	0,9	0,5	0,2
		C	15	3,1	0,17	1,2	1,3	0,2
P. oocarpa	18	M	131	1,1	0,07	0,06	0,06	0,2
		C	19	2,6	0,17	1,1	0,7	0,3
P. elliottii	24		368	0,9	0,10	0,3	0,1	0,1
		M C	65	1,7	0,15	0,6	0,1	0,1
Sugarcane		Stalks	100	1,3	0,08	1,1	0,1	0,2
Coffee tree		Pulp	2	16,5	1,50	26,0	3,5	1,5
Maize		Grains	5	23,0	5,60	7,0	0,4	2,0
Soya		Grains	3	66,7	8,67	19,0	3,3	3,3

(*) Tree components: M=wood; C=Bark. Source: GONÇALVES (1995).

TABLE 12 - Nutrient concentration in the different components of *Eucalyptus saligna* Smith. trees at 7 years of age (1 cutting cycle)

COMPONENT -	NUTRIENTS (%)					
	N	P	K	Ca	Mg	S
SAND	0,07	0,02	0,11	0,10	0,03	0,01
CASCA	0,21	0,09	0,38	3,34	0,39	0,05
RAMOS	0,19	0,04	0,34	0,90	0,15	0,02
SHEETS	0,92	0,11	1,01	0,84	0,26	0,13

Source: VITAL et al. (1999).

TABLE 13 - Tree phytomass and stored nutrients (kg/ha) in the different components of *Eucalyptus saligna* Smith and their respective percentages of the total weight

COMPONENT	UNIT	BIOMASS	ELEMENTS					
			N	P	K	CA	MG	S
SAND	Kg/ha	145180,48	102	29	160	145	44	15
	%	81,5%	50,63%	55,04%	51,8%	18,24%	32,71%	47,5%
CASCA	Kg/ha	14657,40	31	13	56	490	57	7
	%	8,2%	15,33%	25,00%	18,07%	61,5%	42,94%	23,98%
RAMOS	Kg/ha	13615,25	26	6	46	123	20	3
	%	7,6%	12,89%	10,33%	15,01%	15,39%	15,34%	8,9%
SHEETS	Kg/ha	4616,18	43	5	47	39	12	6
	%	2,6%	21,16%	9,63%	15,12%	4,87%	9,01%	19,63%
TOTAL	Kg/ha %	178069,31 100	200,75	52,76	308,31	796,06	133,13	30,57

Source: VITAL et al. (1999).

The quantities listed in Table 12 and Table 13 give an indication of how much of each nutrient, in the form of fertiliser, should be returned in a cutting cycle to avoid depleting the soil. It should be borne in mind that many nutrients are replenished by rainfall, that the continuous weathering of the soil itself makes nutrients available and that the quantities exported vary according to the species used, the parts of the trees that are harvested and the region considered (soil, climate, topography). It is therefore advisable to carry out a complete study of the nutrient balance for each situation before making any decisions, including defining the degree of impact caused to soil fertility by eucalyptus cultivation.

4.2 Allelopathy

SCARPINELLA (2002) defines allelopathy as the interference of one plant on another through substances produced by the former and released into the environment, such as phenolic compounds and terpenes. The same author and LIMA (1993), cite various authors who state that some eucalyptus species have allelopathic substances, which can be released by volatilisation, leaching and exudation, as well as by the decomposition of organic matter, resulting in the retardation of the growth of other species and point out that the references found of eucalyptus species that have allelopathic substances are as follows: *E. camaldulensis*, *E. globulus, E. rostrata, E. pilularis* and *E. microteca*. No references were found for the species most commonly used for pulp production in Brazil.[8]

4.3 Effects on water resources

The effects of forests on water resources have been considered beneficial, but production activities and infrastructure create negative impacts.

The main beneficial effects of forests on water resources, according to IPEF (2003), ROCHA (2001) and ROCHA & KURTZ (2001), are:

> forests, as vegetation cover, with their roots and litter, prevent the direct impact of raindrops on the soil, allow the water to infiltrate gradually and prevent the surface water from running off freely, slowing down the rate of runoff, thereby all, reduce the process of natural soil erosion, which can lead to an overload of nutrients in natural water bodies - according to ROCHA (2001), the average infiltration rate of

[8] Note: In the eucalyptus plantations in the south of Bahia and north of Espírito Santo, it was observed that the undergrowth only reduces the intensity of competition to the eucalyptus when the canopy closes the tree cover, which occurs 3 to 4 years after planting, which shows that if there is allelopathy, at least some species are tolerant.

rainwater in forest soils is 150mm/hour, while in pasture areas it is 12mm/hour and in agricultural areas it is 6mm/hour;

> forests promote the formation of leaf litter and impregnate the soil with organic matter. As organic matter absorbs a lot of water, it acts as a water store in the soil, gradually releasing the absorbed water, thus improving the water regime (ROCHA & KURTZ, 2001);

> Forests act as thermal insulators, according to ROCHA & KURTZ (2001), and as a barrier to excess insolation underneath themselves, reducing temperature variations in the soil and water, facilitating the development of the biota that lives under the tree cover and in the bodies of water that receive its influence;

> terrestrial forest biota produce food for aquatic biota, promoting their development in water bodies.

On the other hand, there are the negative impacts of monocultures, infrastructure and forestry production activities, as can be deduced from some studies such as those carried out by IPEF and the study by CÂMARA & LIMA (1999); the main ones are listed below:

> interception of rainwater by the canopy, facilitating evaporation and reducing the amount of water that reaches the ground;

> water consumption (transpiration);

> release of excess nutrients carried by rainfall and consequent eutrophication of water bodies;

> water consumption in irrigation operations and dilution of agrochemicals;

> road erosion (studied in a separate section in this chapter);

> water pollution from packaging, metals, fuels and lubricants used in forestry operations;

> use of agrochemicals in forestry that are carried into water bodies with the rain (fertilisers and agrochemicals).

Most of these effects, although known, have only recently begun to be studied through research and experiments in eucalyptus-growing areas. Among the studies that have been carried out is the programme started in 1989 by IPEF, which, in 2001, signed an agreement with various organisations, creating a micro-watershed monitoring network to clarify many doubts about the effects on water resources of both natural and planted forests, of which some results stand out:

Results showed that there was no impact of harvesting on water production in the micro-basin monitored in Copener (now Bracell) forest areas. In the general evaluation of the results of the monitoring of chemical variables, no significant changes in water quality were identified. In 2001, the company began monitoring biological indicators of the aquatic ecosystem.

Klabin-Riocell's forested experimental micro-basins in Guaíba-RS show more stable hydrological functioning than the micro-basin with pasture in terms of water balance and flow regime, characterised by greater infiltration, lower peak direct runoff and lower hydrological response (direct runoff/precipitation ratio). On the other hand, analysing the hydrological response of the eucalyptus micro-basin during forest harvesting operations showed that there were no significant changes. Also in this same project, comparing the two eucalyptus watersheds in average terms for the period before felling and the first year after clear-cutting, no drastic changes caused by forest harvesting were found in relation to the water quality parameters analysed.

In terms of the water balance, the results were consistent, with a slight downward trend in rainfall being observed in the three years before felling, as well as an increase in rainfall during the first year after clear-cutting, results that are in line with the trends observed in several other similar studies.

Based on what IPEF has said, it can be deduced that forests and forestry crops have less impact than pastures, and have greater advantages during periods when the tree cover is closed; on the other hand, clear-cutting the forest, used in eucalyptus cultivation for pulp, has significant impacts on water quality, causing an increase in the amount of sediment, changes in colour and an increase in the turbidity of watercourses, according to CÂMARA (1999).

4.4 Water consumption

The effect of eucalyptus forests on soil water is similar to that of other highly productive crops, with some advantages over annual crops, according to REVISTA DA MADEIRA (Sept 2001). *Eucalyptus* species grown on Brazilian soil are as efficient at saving the water they remove from the soil as other forest species and more efficient than most annual crops (see example in Table 14). According to SCARPINELLA (2002), planting eucalyptus does not increase rainfall, but neither does it lead to desertification; its action on the soil is beneficial, reducing the erosion process, improving infiltration conditions, water storage and the soil's chemical and physical properties.

TABLE 14 - Water use efficiency of some species

SPECIES	KG OF BIOMASS / KG OF WATER CONSUMED
Pongamia pinnata	0,8
Prosopis juliflora	1,7
Albizzia lebbek	1,7
Eucalyptus tereticornis	1,9
wheat	0,98
beans	0,5
maize	1,08
potato	0,6

Source: REVISTA DA MADEIRA (May/2002).

4.5 Biodiversity

Monocultures always lead to a reduction in biological diversity; be it eucalyptus or any other crop, according to LIMA (1993).

Fauna habitat conditions can be improved with appropriate forest management practices. Many of the large pulp companies exceed the requirements of environmental legislation by adopting programmes, based on international principles of environmental management, which include the conservation of natural resources under much stricter conditions than those required by Brazilian legislation, involving, among other actions, the creation of corridors for native biota, maintaining mosaics of cultivation plots of different ages, building dams and planting fruit trees along the areas, retaining natural forest reserves undisturbed and keeping some areas unplanted (as there are species that depend on open areas for breeding), according to REVISTA DA MADEIRA (Sept. /2001)./2001).

4.6 Use of pesticides

Pesticides, according to Federal Law No. 7.802 of 11 July 1989, are:

Products and components of physical, chemical or biological processes intended for use in the production, storage and processing of agricultural products, in pastures, in the protection of native or established forests and other ecosystems and also in urban, water and industrial environments, whose purpose is to alter the composition of flora and fauna in order to preserve it from the harmful action of living beings considered harmful, as well as substances and products used as defoliants, desiccants, growth stimulators and inhibitors.

According to KRÜGNER (1980), LIMA (1993) and SILVEIRA *et al.* (2001), among the problems for eucalyptus cultivation in Brazil are leaf-cutting ants, *Atta spp.* and *Acromyrmex spp.*, respectively known as saúva and quenquém, which are mainly controlled with formicide baits and the application of toxic chemical products used directly in the nests, in powder, liquid or nebulisable liquid formulations; other pests and diseases are:

> canker (Cryphonectria cubensis and Valsa ceratosperma) - controlled mainly through genetic selection; boron fertilisation increases plant resistance;

> rust (Puccinia psidii) - controlled mainly through genetic selection;

> Powdery mildew (Oidium sp.) - more common in nurseries, controlled with benomyl;

> grey mould (Botrytis cinerea) - most common in nurseries, controlled through management techniques and chemical control with sprays of thiram, manzate, captan, iprodione or vinclozolin;

> caterpillars (Lepidoptera, Geometridae: Thyrinteina arnobia and Glena spp. and Lepidoptera, Riodinidae: Euselasia sp.) - can generally be controlled using biological agents such as bacilli and viruses, by genetic selection and using the following natural enemies: Deopalpus sp. (Diptera, Tachinidae), parasitises caterpillars and emerges from pupae; Winthemya sp. (Diptera, Tachinidae), parasitises caterpillars and emerges from pupae; Patelloa similis (Diptera, Tachinidae), parasitises caterpillars and emerges from pupae;

Euphorocera sp. (Diptera, Tachinidae), parasitises caterpillars and emerges from pupae; Archytas sp. (Diptera, Tachinidae), parasitises caterpillars and emerges from pupae; Tetrastichus sp. (Hymenoptera, Eulophidae), parasitises pupae; Apateticus sp. (Hemiptera, Pentatomidae), a predator of caterpillars, pupae and adults;

Alcaeorrhynchus grandis (Hemiptera, Pentatomidae), a predator of caterpillars.

> termites (Isoptera: Kalotermitidae, Rhinotermitidae and Termitidae) - controlled using long-residual insecticides with the active ingredients Fipronil or Bifenthrin;

> beetles (Coleoptera, Crysomelidae, Costalimaita ferruginea vulgata, yellow beetle) - controlled by hand-picking and phosphorus insecticides;

> aphids and thrips - have increased considerably in eucalyptus plantations in recent years and can be combated with various chemical insecticides such as some phosphates;

> leaf spots (Cylindrocladium candelabrum, C. ilicicola, C. parasiticum, C. pteridis and C. quinqueseptatum) - common in seedling nurseries and eucalyptus plantations, but the damage is not considerable, the disease must be controlled by genetic selection and application of cupric fungicides or dithiocarbamates, alternated with benomyl;

> seedling damping-off (Cylindrocladium candelabrum, C. clavatum, Rhizoctonia solani, Pythium spp, Phytophthora spp. and Fusarium spp.) - appropriate management techniques and the use of fungicides for the species are recommended for control: Pythium and Phytophthora the use of metalaxyl is recommended; Rhizoctonia, a combination of captan with a dithiocarbamate (maneb, zineb or thiram); Cylindrocladium or Fusarium you can use benomyl with captan or thiram; you can disinfect the substrate beforehand using water vapour at a temperature of between 80 and 90°C for around 7 to 8 hours, or drying in the sun;

> root rot (Phytophthora and Pythium) - usually caused by poor drainage of the substrate, it can be combated in a similar way to tipping over;

> rot of cuttings and mini-cuttings (Cylindrocladium spp., Rhizoctonia solani, Fusarium spp., Botryosphaeria ribis and Colletotrichum sp) - controlled with the use of sodium hypochlorite and fungicides to disinfect

all the materials used in the production of cuttings and the greenhouse, such as copper sulphate; sometimes the disease is caused by calcium deficiency.

> nematodes - attack the roots of plants, which in nursery conditions can be controlled with nematicides or fumigants.

Other agrochemicals used in eucalyptus cultivation are pre and post-emergent herbicides, with the aim of reducing competition.

Table 15 shows the characteristics of some pesticides used in eucalyptus cultivation.

TABLE 15 - Characteristics of some pesticides used in eucalyptus cultivation

Class	Product	target	Active ingredient	DL50 Oral (mg/Kg)	DL50 Dermal (mg/Kg)
Formicide	ATTAMEX	Leafcutter ants	Sulfluramide	3250 (mice)	>2500 (rabbits)
Formicide	DURSBAN	Termites and leaf-cutting ants	orgophosphorus	>201 (mice)	>1900 (rats)
Cupinicide	PYRETHRIN	Termites	Cyfluthrin	>800 (rats)	>5000 (rats)
Insecticide	DECIS	Caterpillars	Deltamethrin	>1190 (rats)	>12000 (rats)
Herbicide	GlyphosateNortox- NA	Mato competition	Glyphosate	>6000 (rats)	>12000 (rats)
Herbicide	CHOPPER-NA	Eucalyptus sprouting	Imazapyr	5000 (mice)	>2000 (rabbits)
Fungicide	BATFIDAN-CA	Fungi	Triadimene	1700 (mice)	>5000 (rats)
Fungicide	Benlate	Various fungi	Benomyl	>10000 (rats)	>10000 (rabbits)

Source: BAUCH *et al.* (2001).

Many problems with pesticides have been minimised through monitoring, drastically reducing the use of toxic substances to control pests, diseases and weeds, especially ants and matocompetition, according to FLORIANO (1998).

4.7 Impacts of roads

The total area owned by the largest eucalyptus pulp and paper companies is around 3,727,000 hectares, of which approximately 483,000 hectares are classified as road areas, infrastructure, water bodies and other areas not used for forestry, i.e. around 12.96 per cent of the properties.

Roads, or motorways, are complex structures whose main purpose is to serve as a land transport route for people and cargo.

On the other hand, roads represent one of civilisation's greatest evils in terms of their impact on the natural environment; they also cause problems for the human environment itself.

A motorway can be classified as an engineering project made up of a roadway and engineering structures. Its impacts start with planning, continue through the implementation and construction phases, right up to the operational phase, when the quality of its maintenance has major implications. The environmental impact assessment of motorways should include all phases, but in Brazil it is still in its infancy in the operational phase, with little legislation requiring it in this phase, although it perhaps creates more impacts than in the others, as it has an indefinite lifespan, which could be centuries.

From a legal point of view, highways must be the subject of an EIA/RIMA whenever they have two or more lanes, which includes all federal highways, most state highways and some municipal highways. Federal legislation on motorways is mainly represented by the following acts:

> Law No. 10.233, of 5 June 2001 - Provides for the restructuring of waterway and land transport, creates the National Council for the Integration of Transport Policies, the National Land Transport Agency, the National Waterway Transport Agency and the National Department of Transport Infrastructure, and makes other provisions;

> CONAMA Resolution No. 001 of 23 January 1986 - Establishes the definitions, responsibilities, basic criteria and general guidelines for the use and implementation of Environmental Impact Assessment as one of the instruments of the National Environmental Policy;

> CONAMA Resolution No. 237, of 19 December 1997 - Establishes procedures and criteria for environmental licensing, in order to effectively use the licensing system as an environmental management instrument, established by the National Environmental Policy;

> Decree No. 96.044, of 18 May 1988 - Approves the Regulations for the Road Transport of Dangerous Products, and makes other provisions.

In addition to these, it is necessary to base oneself on the rest of the environmental legislation and that which deals with mineral extraction activity for borrow areas, quarries, limestone, sand, macadam, gravel and clay exploitation, represented mainly by the following acts: CONAMA Resolution No. 009 of 06/12/1990; CONAMA Resolution No. 010 of 06/12/1990; Federal Law No. 4.771 of 15/09/1965; Federal Law No. 9.985 of 18/07/2000; Decree Law No. 227 of 27/02/1967; Federal Law No. 9.314 of 14/11/1996; Federal Decree No. 3.358 of 02/02/2000; Ministerial Order No. 23 of 03/02/2000.

NEARY and HORNBECK *apud* SOUZA and SEIXAS (2001) estimate that more than 90 per cent of sediment production in forests comes from the construction and maintenance of forest harvesting roads. The same authors, SOUZA and SEIXAS (2001), studying the cover of cut and fill slopes in Jacareí, SP, found sediment production values of 0.67 t/ha in six months (July-November 2000) after stabilising the cover with batatais grass on 95% of the surface, with rainfall over the 12 months of the study of 1122 mm and slope gradients of 85%.

In forestry, the main purpose of roads is to provide access to production areas and they are planned to result in the lowest cost for harvesting and transporting timber, operations where the greatest costs of silvicultural activity are concentrated. Their density varies from 15 to 100 metres per hectare, with the size of plots being designed to result in a ratio between length and width generally close to 2:1, according to DURLO (1979).

Sediment production by cut and fill slopes with different road covers is shown in Table 16.

The smallest and most typical forest management unit surrounded by a road is the stand. In forestry, plots are planned to be of such a size that they provide the lowest transport costs (removal of timber to the road). The average size of these units used by companies is around 25 hectares. The ideal surface area and ratio between length and width varies according to topography, soil type and forest productivity, but from the information mentioned above it can be inferred that the average road density is around 60 linear metres per hectare. If the average road width is around 5.5 metres, then around 3.3% of the surface area of forest properties is represented by roads.

TABLE 16 - Sediment production by cut and fill slopes with different coverings.

PERIOD	WITNESS[3]	GRAMAb	CASCAc	TELAd
1 - 21 Jan/99	7,09	0,24	0,16	-
21 Jan - 2 Feb	3,19	2,96	0,01	-
2 - 8 Feb	16,55	0,53	0,02	-
8 - 12 Feb	11,86	0,68	0,03	-
8 - 23 Feb	3,48	0,85	0,31	-
23 Feb - 2 Mar	4,18	1,15	0,06	-
2 - 18 Mar	0,14	0,01	0,01	0,10
18 Mar - 1 Apr	1,47	0,03	0,02	0,10
1 - 20 Apr	0,08	0,12	0,04	0,10
20 Apr - 18 May	0,63	0,39	0,04	0,55
18 May - 9 Jun	0,14	0,03	0,01	0,02
9 Jun - 13 Jul	0,31	0,05	0,04	0,10
13 Jul - 2 Sep	0,42	0,14	0,07	0,40
2 - 23 Sep	1,24	0,13	0,02	0,20
23 Sep - 18 Oct	0,47	0,14	0,02	0,20
18 Oct - 26 Nov	0,56	0,06	0,02	0,40
26 Nov - 5 Jan/00	1,26	0,15	0,10	0,10
Subtotal	6,72	1,25	0,39	2,27
Total	53,07	7,66	0,98	2,27

(a) Witness: slope without cover (bare soil). (b) Grass: slope covered with "Batatais" grass slabs. (*Paspalum notatum* Flügge).

(c) Bark: slope covered with chopped *Eucalyptus spp*. bark (waste from the pulp and paper industry). (d) Screen: slope covered with organic material impregnated with grass seeds. Source: SOUZA and SEIXAS (2001).

Extrapolating the aforementioned data, using an average sediment production value for grass-stabilised slopes of 1.34 t/ha/year, over 1 million hectares, with 3.3% roads, would result in a sediment production of close to 44,000 tonnes per year, without taking into account that roadbeds produce much more sediment than grass-stabilised slopes. As the minimum values and the best situations were used to make the estimate, this is an idea of the minimum impact caused by forest roads in the Brazilian eucalyptus pulp and paper sector.

In addition to the problems mentioned above, roads are harmful to wildlife, which is constantly run over and killed, making them one of the biggest threats to the lives of wild animals, with more than 1 million animals run over every day in Brazil.

4.8 Other impacts

In addition to the impacts mentioned above, many others are associated with Brazil's production of *Eucalyptus* wood for pulp:

> soil compaction by machinery and equipment;

> infrastructure works;

> pollution by waste and effluents from human activities and product packaging;

> soil pollution from machinery, vehicle and equipment maintenance activities;

> air pollution and noise pollution caused by machinery and vehicles;

> pollution and explosions caused by accidents with hazardous materials in their transport and storage;

> human and animal poisoning caused by accidents involving the use of toxic substances;

> Use of transgenic *Eucalyptus*;

> Barrage, derivation and consumption of water for firefighting, for irrigation of plantations and for dilution and application of agrochemicals.

CHAPTER 5

5 POLICIES, LEGISLATION, CERTIFICATION AND AIA

Environmental management policies, including legislation and certification process standards, are covered to demonstrate the origin and objectives of Environmental Impact Assessment (EIA) policies and methods.

5.1 Introduction to environmental management policies

"Forests must be utilised on the basis of their natural characteristics for the well-being of the population, managed and conserved with care and with the responsibility of leaving a good legacy for future generations." *(by CARLOWITZ* apud *GROBER*)

The concept of sustainability, mentioned above, is the first of all known and documented concepts; it was issued by CARLOWITZ in his book "Sylvicultura Oeconomica" in 1713. In general terms and based on this, environmental management can be understood as the administration of environmental resources with the aim of conserving them and ensuring that future generations find an environment compatible with their living needs.

When talking about environmental management policies, it is necessary to distinguish between their character and level of scope.

Environmental management policies can be categorised as public or private.

The Constitution of the Federative Republic of Brazil establishes the guidelines for the development of national public environmental policies in Chapter VI, Article 225; in addition to this article, the Federal Constitution also establishes guidelines with environmental implications in Articles 5, 23, 24, 129, 170, 174, 187, 186 and 220.

The concept of public environmental management, quoted during the Seminar on the Training of Educators to Work in the Environmental Management Process in Brasilia in 1995, emphasises the conciliatory aspect of the state with regard to environmental issues, and is reproduced below:

Public environmental management is a process of mediating interests and conflicts between social actors who act on the physical, natural and built environments. This mediation process continually defines and redefines the way in which the different social actors, through their practices, alter the quality of the environment and also how the costs and benefits resulting from the actions of these agents are distributed in society.

According to this concept, which still seems appropriate today, with the guidelines of Agenda 21 and the Rio Declaration and CARLOWITZ's concept of sustainability, it is understood that public environmental management policies should aim not only to manage resources to protect the natural environment, but above all to serve as a guide in resolving social conflicts involving environmental issues, with a view to social welfare and the conservation of resources for future generations.

Private environmental policy can be understood as an organisation's statement setting out its intentions and principles in relation to its overall environmental performance, which provides a framework for action and the definition of its environmental objectives and targets. Private environmental management is dealt with extensively in the ISO 14000 series standards, from which the following concept can be deduced: private environmental management is an integral part of an organisation's overall management system and constitutes an administrative, dynamic and interactive process of resources, the purpose of which is to balance environmental protection and pollution prevention

to meet socio-economic needs and help protect human health, through the formulation of a policy and objectives that take into account legal requirements and information regarding significant environmental impacts, with a view to continuous improvement in the organisation's environmental performance in order to meet the needs of a wide range of interested parties and the growing needs of society for environmental protection.

In territorial terms, environmental management policies can be categorised as rural and urban at global level, national level, large regions national level, state level, municipal level and, more recently, at the level of state regions, river basins and ecoregions (ecosystems).

International environmental management policies have been created mainly during the various United Nations conferences related to the environment. The best-known document with the greatest repercussions is Agenda 21, which contains the main environmental and development policies at international level; this document was created during the United Nations Conference on Environment and Development in Rio de Janeiro in 1992, ECO-92. Another document issued at the same conference is the Rio Declaration on Environment and Development, which contains 27 principles for protecting the integrity of global environmental systems and for sustainable human development. The work of large NGOs (such as the World Wildlife Fund - WWF) and the recommendations of the World Bank are important not only at international level, but also in the formulation of Brazilian environmental policies, according to SOUZA (2001). Considering the private aspect of global environmental issues, the policies created by ISO[9] and the World Trade Organisation (WTO) must be taken into account: the former with the ISO 14000 series of international standards, which guide organisations in creating environmental management systems around the world; and the WTO with its restrictions on international trade in endangered species.

In Brazil there is a structure called the National Environmental System (SISNAMA), led by its highest body, the Government Council, which has the function of advising the President of the Republic on the formulation of national policy and government environmental guidelines, with the participation of the Ministry of the Environment and the Legal Amazon as coordinator of the National Environmental Policy Environment, expressed in Federal Law No. 6938/81, with CONAMA - the National

---

[9] ISO - International Organisation for Standardization.

Environmental Council - as its advisory and deliberative body and IBAMA - the National Institute for the Environment and Renewable Natural Resources - as its executive body. In the states, the state environmental secretariats do the coordinating work, the state environmental councils are the advisory and deliberative bodies and the executive bodies have generally been created as foundations or public companies that provide services to the direct administration, such as FATMA [1011] in Santa Catarina, FEPAM[11] in Rio Grande do Sul and CETESB[12] in São Paulo; in some states the executive body is a department linked to the environmental secretariat. At municipal level, which still doesn't exist in many municipalities, there is a municipal secretariat responsible for coordinating municipal environmental policy, an environmental council as an advisory and deliberative body and an executive body, the latter in many cases being the same as the state body, through agreements signed between municipalities and states, due to the difficulty and cost of setting up a municipal body of this nature for many Brazilian municipalities.

In water resources management, the territorial administrative unit is the river basin, by virtue of Law 9.433/97. The division into basins is carried out in terms of large national basins and these, by state, are divided into sub-basins that can still be divided and managed by micro-basins, but the main administrative units are the national territory as a whole, the large basins and the sub-basins, the first two at federal level and the latter at state level.

According to Muller (1997), ecosystem management is a recent trend in terms of natural environment protection policies; Ecosystem management is about applying the new knowledge developed in

---

[10] FATMA - Foundation for Technology and the Environment - SC.

[11] FEPAM - Henrique Luis Roessler State Foundation for Environmental Protection - RS.

[12] CETESB - Environmental Sanitation Technology Company - SP.

environmental sciences to the management of regions, according to their physiognomy, in order to maintain their richness and, above all, to maintain ecological functions and structures both at the level of species and their habitats, as well as at the level of the larger ecosystems that generate life, maintaining the flow of nutrients and structures, such as areas of large and old trees, wetlands and pastures, which create different systems in the landscape and are the key to maintaining biodiversity. In this respect, we can distinguish between the levels of large biomes, ecoregions within biomes, regional physiognomies within ecoregions, local ecosystems within regional physiognomies and species habitats within local ecosystems. From this, it can be deduced that the main objective of ecosystem management is the conservation of biodiversity and ecosystem productivity and that it can be both public and private, affecting territories at different levels. For example, when it comes to the habitat of an endemic species that is entirely within the area of a sub-basin within the boundaries of a single municipality, there is minimal territory; for the Amazon biome, the territory considered is huge and affects several countries. Brazil seems to be taking this new direction, as IBAMA recently finalised its inventory of Brazilian ecosystems and has expressed the government's intention to adopt differentiated environmental management policies according to the characteristics of each physiognomy. This seems to be the right thing to do, and it should begin by issuing a forestry code based on the new policies, eliminating the generalised restrictions and permissions for the country as a whole, present in the 1965 code and the acts that modified it, which turn every Brazilian rural worker into an environmental criminal and the civil servants responsible for enforcing the law into criminals for prevarication, leaving a few questions: who is right, the law or the population? should the law help solve problems, or not? should the law prevent social problems from being created for the majority of the population, or not? - among others that may arise when thinking about ways of applying Law 4771/65 with its incompatible requirements for the majority of environmental situations (anthropogenic and natural) found in Brazilian territory and maintained in the new law Law 12.651, of 25 May 2012, which established rules for the protection of native vegetation in areas of permanent preservation, legal reserve, restricted use, forest exploitation and related matters.

The functions and attributions of the public authorities in implementing the National Environmental Policy are listed in Decree No. 99.274 of 06/06/1990, which regulates Law No. 6.902 of 27 April 1981 and Law No. 6.938 of 31 August 1981, which respectively provide for the Creation of Ecological Stations and Environmental Protection Areas and the National Environmental Policy.

TABLE 17 - Biodiversity and endemism in Brazil

ANIMALS	NO. OF SPECIES	% ENDEMISM	TOTAL % of SPECIES WORLDWIDE	WORLD RANKING
Freshwater fish	>3000	-	22	1
Amphibians	517	57	12	2
Birds	1622	>12	9	2
Mammals	524	25	9	1
Reptiles	468	37	7	5

(*) Source: BERNARDES (2000) .

5.2 Principles of environmental management

Principles, in a simplistic view, are basic action policies. They are the guidelines from which all secondary actions and policies stem. There are some differences between public and private environmental management principles, but both are formulated out of a need to solve environmental problems that affect society, whether out of economic, social or cultural interest.

International environmental management, as mentioned above, is mainly based on the principles of the Rio Declaration[13] , Agenda 21[14] , the

[13] Rio Declaration on Environment and Development. United Nations Conference on Environment and Development, UN, Rio de Janeiro 1992. **Proceedings...** [Brasília]: IBAMA, n.d. Available at: < http://www.ibama.gov.br > . Accessed on: 12 August 2002

[14] AGENDA 21. United Nations Conference on Environment and Development, UN, Rio de Janeiro

precepts dictated by international organisations such as the UN, ISO and WTO, and can also include the requirements of international financial organisations such as the World Bank and the Inter-American Development Bank.

The Federal Constitution states that ***"Everyone has the right to an ecologically balanced environment"***, and this is the first of all the principles governing environmental policy in the country. Any act or action contrary to this basic principle is illegal. Other principles established in the Constitution are: **sustainability**, which imposes on public authorities and the community the duty to defend and preserve the environment for present and future generations; and **environmental responsibility**, which places the onus for recovering from environmental impacts and damage on the agent who caused them. The other basic principles of Brazilian public environmental management, derived from the first three, are established in Law 6938/81, in its Article 2, as follows:

Art. 2 - The National Environmental Policy aims to preserve, improve and restore environmental quality favourable to life, in order to ensure conditions for socio-economic development in the country, the interests of national security and the protection of the dignity of human life, taking into account the following principles:

I - government action to maintain ecological balance, considering the environment as a public heritage to be necessarily ensured and protected, with a view to collective use;

II - rationalising the use of soil, subsoil, water and air;

III - planning and monitoring the use of environmental resources;

IV - protection of ecosystems, with the preservation of representative areas;

V - control and zoning of potentially or effectively polluting activities;

VI - incentives for the study and research of technologies geared towards the rational use and protection of environmental resources;

VII - monitoring the state of environmental quality;

VIII - recovery of degraded areas;

IX - protection of areas threatened with degradation;

X - environmental education at all levels of education, including community education, with the aim of enabling the community to actively participate in the defence of the environment.

In the private sector, environmental management is a factor in commercial competition due to the tendency of most people to favour environmentally friendly products and services. In addition, international

---

1992. **Proceedings...** [Brasília]: IBAMA, n.d. Available at: < http://www.ibama.gov.br > . Accessed on: 12 August 2002

trade contracts are made easier for organisations that adopt management systems based on internationally recognised standards such as those of ISO and the FSC.

Private environmental management, according to the ISO 14000 series standards (ABNT, 1996/1996a), in the exclusively business sphere, has among its principles, but is not limited to these:

- Recognise that environmental management is among the organisation's highest priorities;
- Establish and maintain communication with internal and external stakeholders;
- Determine the applicable legal requirements and the environmental aspects associated with the organisation's activities, products or services;
- Developing the commitment of management and employees to protecting the environment, with a clear definition of responsibilities and accountability;
- Encourage environmental planning throughout the life cycle of the product or process;
- Establish a process to achieve the target performance levels;
- Provide appropriate and sufficient resources, including training, to achieve the targeted performance levels on an ongoing basis;
- Evaluating environmental performance in relation to the organisation's environmental policy, objectives and targets, seeking improvements where appropriate;
- Establish a management process to audit and critically analyse the EMS and to identify opportunities for improving the system and the resulting environmental performance;
- Encourage service providers and suppliers to establish an EMS.

5.3 Instruments of the National Environmental Policy (PNMA)

To summarise, it can be said that the PNMA uses five main instruments (supported by tools, systems and methodologies):

> Licensing - is based on Environmental Impact Assessment (EIA), Environmental Impact Study (EIS), Environmental Impact Report (RIMA), Environmental Control Plan (PCA), Degraded Area Recovery Plan (PRAD), Environmental Assessment Report (RAA), Environmental Control Report (RCA), Risk Analysis (AR), Environmental Feasibility Study (EVA), Basic Environmental Project (PBA), Term of Reference (TR), Public Hearing (AP), establishment of environmental quality standards and information system and federal environmental technical register;

> Economic incentives - the National Environmental Fund (FNMA), funding from the BNDES for large projects, etc;

> Economic inhibitions - Ecological taxes (ICM, Forest Restoration Fee, etc);

- Punishment - Environmental Crimes Law, Forest Code, etc;
- Conservation - National System of Conservation Units (SNUC), Ecological Corridors, Pilot Programme for the Protection of Brazil's Tropical Forests (PPG7), restrictions on the use of natural resources on private properties (imposed by Law 4771/65, among others), environmental recovery, self-supply and forest replacement.

5.4 Private environmental management instruments

In the private sector, the main instruments included as requirements in the ISO 14000 series standards (ABNT, 1996/1996a) are the following:

- Commitment from the organisation through the quality policy;
- Planning;
- Environmental impact assessment;
- Environmental performance assessment;
- Continuous improvement process (example: PDCA[15]);
- Quality system documentation;

[15] Note: PDCA (Plan, Do, Control, Act) - a continuous improvement process created by Shewhart and publicised by Deming, according to MIRSHAWKA (1990).

- Records of operational and environmental activities and elements;
- Training;
- Communication;
- Monitoring;
- Environmental auditing;
- Critical analysis by management.

It would be redundant to explain the environmental management instruments according to the ISO standards, as they are perfectly broken down and explained in the text of the standards themselves.

According to MONTARDO (2002), the environmental management tools for projects are as follows: environmental impact assessment, recovery of degraded areas, environmental monitoring, environmental auditing, environmental risk analysis, investigation of environmental liabilities, environmental insurance and environmental management systems.

It can be deduced that many of the instruments used in public and private environmental management are common, with public management emphasising instruments that lead to environmental protection and control, while private management favours planning and continuous improvement from a current situation.

The main instruments of Brazil's public environmental management policy are, on the one hand, command and control through licensing, which seeks to keep the effects of anthropogenic activities under control and, on the other, conservation through nature conservation units and biota corridors[15] which seek to conserve the most significant parts of the natural environment and human culture in the national territory.

5.5 Environmental Impact Assessment (EIA)

The Environmental Impact Assessment (EIA) was institutionalised in Brazil on 31 August 1981 by Federal Law 6.938 and its application was regulated by CONAMA Resolution 001 of 23 September 1986.

According to TAVARES JÚNIOR (1997), the EIA was introduced in Brazil at the request of international funding bodies and later incorporated as an instrument of the National Environmental Policy, aimed at licensing projects, based on legislation in the United States, which was the first country to require an EIA for government projects, programmes and activities in the late 1960s.

The EIA is the part of the Environmental Impact Study and respective Environmental Impact Report (EIA/RIMA) where the possible impacts of a project, their causes and effects are identified, qualified and quantified.

According to PIMENTEL (1992) *apud* TAVARES JÚNIOR (1997)[(2)], EIA methods can be grouped into two large groups, according to the activity for which their use is most appropriate:

1 - Methods for the identification and summarisation phase - This aims to identify, qualify, quantify and present the probable impacts resulting from a project;

---

[15] Note: Ecological corridors are intended to facilitate the transfer of genes between protected areas and can be considered as conservation units, but are not defined as such by Law 9.985 of 18/07/2000, which established the National System of Conservation Units.

2 - Methods for the evaluation phase - These are used to compare alternatives, and can be categorised according to how opinion is incorporated into the process.

The Environmental Impact Assessment (EIA) is the heart of an EIA/RIMA, because it is through the EIA that the validity or otherwise of an undertaking is defined; it is the identification and assessment of impacts in accordance with environmental principles, criteria and indicators, seeking to assign values to them, in a systematic way, in relation to all its environmental aspects and at all stages of the undertaking.

According to Law 6038/81, CONAMA is responsible for establishing criteria and standards relating to the control and maintenance of environmental quality with a view to the rational use of environmental resources, and it can delegate responsibility to the states for matters within its remit and the states, in turn, when it falls within the remit of the municipalities, can transfer responsibility to them.

5.6 Forest certification and EIA

The two main forest certification bodies in Brazil are ISO and FSC. The two adopt very similar procedures and standards. The FSC-Brazil standards can be obtained from the NGO's website. Both systems require compliance with legislation and therefore the Environmental Impact Assessment (EIA) automatically becomes part of the certification programmes, as it is required by Brazilian law.

5.6.1 Forest Stewardship Council (FSC)

The FSC, which stands for Forest Stewardship Council, is a non-governmental organisation founded in 1993 and based in Mexico.

According to REVISTA DA MADEIRA (May 2002), the FSC reached more than 25 million certified hectares in January 2002, which represents around 28% of all certified forest areas in the world, considering the largest forest certification systems in existence, showing faster growth in recent years, which has already reached an average of more than 3 million certified hectares per year, making it the most comprehensive forest certification system in existence; the magazine article also highlights the main characteristics of the FSC programme, which determine its credibility worldwide:

- The first certification system created by individuals and institutions from various countries on all continents;
- It allows NGOs - which have greater credibility with consumers - to act as certifiers.
- It requires a field evaluation;
- Certified forests are monitored annually;
- Certification guarantees the origin of the raw material to the consumer who consumes the products, by certifying the entire production chain (known as the "chain of custody").
- Any member of society can participate in the certification process, and their comments and suggestions are taken into account.
- In the case of forest management certification, a public summary is drawn up and made available.

The principles for FSC forest certification are (FSC, 2024):

Principle 1 - Compliance with the Laws
Principle 2 - Workers' Rights and Working Conditions
Principle 3 - Rights of Indigenous Peoples
Principle 4 - Community Relations
Principle 5 - Forest Benefits
Principle 6 - Environmental Values and Impacts
Principle 7 - Management Planning
Principle 8 - Monitoring and Evaluation
Principle 9 - High Conservation Values
Principle 10 - Implementation of Management Activities

In Brazil, the effort to set up FSC representation dates back to 1994. A Working Group was formalised in 1996 under the coordination of WWF-Brazil. In September 2001, the *Brazilian Forest Management Council* (FSC-Brasil) was founded in Brasilia, with the endorsement of FSC International.

The aim of the FSC is to spread good forest management throughout the world in accordance with Principles and Criteria that reconcile ecological safeguards with social benefits and economic viability. It had issued more than 300 forest management certificates and more than 1,500 chain of custody certificates by 2002.

The FSC label is a guarantee of origin; it certifies that the forest product, or its derivative, comes from a forest that has been managed in an ecologically appropriate, socially just and economically viable way and that complies with all the laws in force.

The benefits of FSC certification are:

Economic - Allows the owner to know exactly how many trees and which species he has and when each one will be available for harvesting. Increases forest yields. Generates competitive advantage. Facilitates access to new markets. Develops and improves the company's public image and the team spirit of its employees.

Environmental - Contributes to the conservation of biodiversity and its associated values: water resources, soils, landscapes and unique and fragile ecosystems. Maintains the ecological functions and integrity of forests. Protects threatened or endangered species and their habitats.

Social - Legalises the activity. Promotes respect for the rights of workers, indigenous peoples and local communities. Contributes to the reduction of accidents at work as a result of the introduction of safety and prevention standards. Improves working conditions. Creates a new space for workers and forest peoples to participate in setting standards and monitoring certified operations. Eliminates forced labour and child labour. Qualifies the labour force.

5.6.2 International Organisation for Standardization (ISO)

ISO is a non-governmental organisation founded in 1947 and based in Geneva. It is an international standardisation forum with 100 members representing their countries, Brazil being represented by ABNT.

According to SOARES (2002a), in 1993 ISO set up Technical Committee 207, responsible for drawing up the ISO 14000 standards, made up of 41 participating countries, 9 observer countries and 13 liaison members (CT207/176).

The number of certificates for environmental management systems in Brazil drawn up until 2001, in accordance with the ISO 14000 standards, is shown in Figure 19 below:

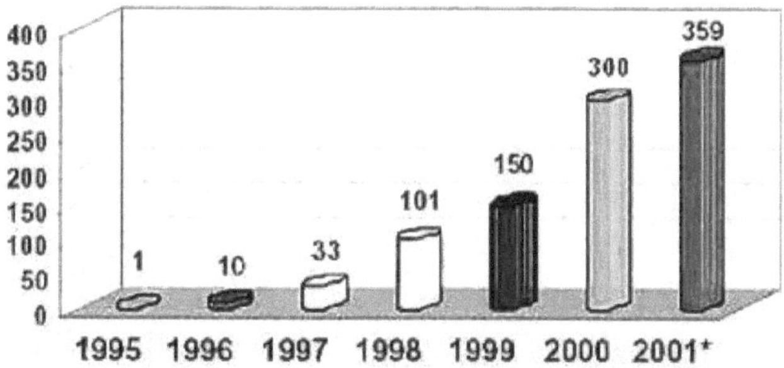

Figure 19 - ISO 14000 certificates issued in Brazil, CHIARADIA (2001).

The principles of an Environmental Management System (EMS) in organisations, according to the ISO 14000 series standards (ABNT, 1996a), are as follows:

Principle 1 - Commitment and policy

It is recommended that an organisation defines its environmental policy and ensures commitment to its EMS.

Principle 2 - Planning

It is recommended that an organisation formulates a plan to comply with its environmental policy.

Principle 3 - Implementation

For effective implementation, it is recommended that an organisation develop the necessary training and support mechanisms to meet its environmental policy, objectives and targets.

Principle 4 - Measurement and evaluation

It is recommended that an organisation measures, monitors and evaluates its environmental performance.

Principle 5 - Critical analysis and improvement

It is recommended that an organisation critically analyses and continually improves its environmental management system, with the aim of improving its overall environmental performance.

With this in mind, the EMS is best seen as an organisational structure, which is recommended to be continuously monitored and periodically critically analysed in order to be able to steer the organisation in response to changing internal and external factors. It is recommended that every person in the organisation accepts their responsibility for environmental improvements.

Private environmental management policy under the ISO standards therefore has the principle of listening to the sectors of society involved, customers, government, suppliers, workers, shareholders, neighbours, etc., to create a system for managing the environmental aspects of their processes and products and improving it continuously.

According to SOARES (2002a), the analysis of environmental impacts in an environmental management system based on the ISO 14000 standards (ABNT, 1996) includes:

List for normal operating conditions (continuous operation of the organisation's equipment).
List for abnormal operating conditions (start-up conditions, stop conditions, speed variations, small foreseeable events, etc).
List for environmental risk conditions (emergencies, incidents, accidents, etc., i.e. imagined and modelled conditions with some probability of occurrence).

The EMS, according to ISO 14000 (ABNT, 1996/1996a), is based on drawing up an environmental policy that will guide the entire system within the organisation, planning, implementation and operation, verification and corrective action, critical analysis by management and continuous improvement . It is applicable to any organisation that voluntarily wishes to implement, maintain and improve an environmental management system, ensure its compliance with its environmental policy and demonstrate this compliance to everyone inside and outside the organisation.

Environmental impact assessment, according to NBR ISO 14004 (ABNT, 1996a), is one of the elements of the EMS and is part of the second principle of NBR ISO 14001 (ABNT, 1996), "Planning", which includes the following elements: identification of environmental aspects and assessment of associated environmental impacts; legal requirements; environmental policy; internal performance criteria; environmental objectives and targets; environmental plans and management programme.

The FSC (Forest Stewardship Council) was set up in February 1994 and launched standards for forest management certification in 1997. The most widespread certification systems in the world are the FSC (Forest Stewardship Council International / Brazil) and the PEFC (Programme for the Endorsement of Forest Certification Schemes). In Brazil, the first actions took place in 1994 and the first area certified by the FSC was in 1995.

Since 2002, CERFLOR (the Brazilian Forest Certification Programme) has been recognised by PEFC and coordinated by INMETRO. Forest management certification is a voluntary process that guarantees that forest management is carried out in a responsible manner, complying with all legislation, in accordance with the principles and criteria of certification, being environmentally appropriate, economically viable and socially just.

The certified area in Brazil is shown in Table 18.

TABLE 18 - Certified area in Brazil, 2022

Product Type	NATIVE		PLANTED		TOTAL	
	№	Area (ha)	№	Area (ha)	№	Area (ha)
		FSC system				
Logs	17	798.467	194	7.129.488	211	7.927.955
Wood for chips or particles	1	33.554	18	244.173	19	277.727
Food	2	150.965	2	19.260	4	170.225
Sawn and processed timber			7	143.357	7	143.357
Gums, oils and derivatives	1	91.556	3	31.064	4	122.621
Other non-timber products	1	91.556			1	91.556
Bark	1	7.074	1	75.176	2	82.250
Wood for charcoal			1	75.176	1	75.176
Plants or their parts	1	33.554	3	37.320	4	70.874
Treated and impregnated wood			5	51.523	5	51.523
Pulp						
Total	24	1.206.728	234	7.806.538	258	9.013.266
		Cerflor/PEFC system				
Logs	1	219.137	22	3.014.524	23	3.233.661
Wood for chips or particles			2	359.719	2	359.719
Wood for charcoal			1	230.617	1	230.617
Sawn and processed timber	1	219.137			1	219.137
Pulp			2	214.165	2	214.165
Bark						
Plants or their parts						
Gums, oils and derivatives						
Food						
Other non-timber products						
Treated and impregnated wood						
TOTAL	2	438.274	27	3.819.025	29	4.257.300

Source: IBÁ (2023).

CHAPTER 6

6 Final considerations

It can be seen that eucalyptus cultivation in Brazil has brought benefits such as jobs and income, improved social and economic indicators, carbon sequestration, soil protection and climate regulation, but it is not free of negative environmental impacts, most of which can be mitigated.

The sector's economic prospects are good for the future, with consumption of eucalyptus products expected to grow, as this is one of the rural crops that causes the least environmental impact and involves much less risk than annual crops and food products in general.

Forestry Engineering was born under the aegis of sustainability some three centuries ago. At the end of the 20th century, forest management certification systems gave an even greater boost to guaranteeing the sustainability of the sector, as well as current legislation aimed at maintaining production, conservation and environmental preservation areas in good condition, affecting the majority of Brazilian eucalyptus plantations.

REFERENCES

ABAF. Laminate flooring. Bahia Association of Forest-Based Companies. Available at: <https://www.abaf.org.br/produtos/pisos- laminates/>. Accessed on: 2024.

ABNT. Interpretation NBR ISO 14001 (1996). Rio de Janeiro: ABNT- Cb-38/Sc-01/Interpretation Group, July 2001.

. NBR 14789: Forest Management - Principles, criteria and indicators for forest plantations. Rio de Janeiro, 2001.

. NBR ISO 14001: Environmental Management Systems - Specification and Guidelines for Use:. Rio de Janeiro, 1996.

. NBR ISO 14004: Environmental management systems - General guidelines on principles, systems and support techniques. Rio de Janeiro, 1996a.

. NBR ISO 14010: Guidelines for environmental auditing - General principles. Rio de Janeiro, 1996b.

AGENDA 21. United Nations Conference on Environment and Development, UN, Rio de Janeiro 1992. Proceedings... [Brasília]: IBAMA, n.d. Available at: < http://www.ibama.gov.br > . Accessed on: 12 August 2002.

ARACRUZ CELULOSE. Contains institutional, financial, social and technical information. Available at: < http://www.aracruz.com.br >. Accessed on: 31/Jan/2003.

ARAÚJO, Ademir S. F de. Biodegradation, extraction and analysis of glyphosate in two types of soil. Dissertation (Master's Degree in Agronomy) - University of São Paulo, Luiz de Queiroz College of Agriculture, Piracicaba, 2002.

b.FOREST. Roundwood price index in Brazil. b.Forest, a.8, n.80, p.60, Jan-Feb/2022.

BACHA, Carlos J. C. Florestas - O sistema agroindustrial da madeira. [Brasília]: Embrapa, 2000. Available at: <http://www.embrapa.br:8080/aplic/rumos.nsf/>. Accessed on: 06/07/2002.

BAHIA SUL. Contains institutional, financial, social and technical information. Available at: < http://www.bahiasul.com.br >. Accessed on: 31/Jan/2003.

BAUCH, Roberto E.; KIKUCHI, Mário Y.; PIRES, José S. R. Evaluation of the recertification of the forest management of the EUCATEX S.A. Ind. e Com. forest plantations in the regions of Salto, Botucatu and Buri in the State of São Paulo. Eucatex, certificate no. SCS-FM/COC-00040P, 1ª ver. Salto: October 2001. Report.

BNDES. The pulp and paper sector in Brazil and worldwide. Rio de Janeiro: 2002.

. Forest Products - Market Pulp. Rio de Janeiro: BNDES, FINAME, BNDESPAR, 2002a.

BRACELPA. Reforestation - History. Available at: <http://www.bracelpa.

org.br/reflorestamento.asp> . Accessed on: 15/Sep/2002.

BRASIL. Ações Setoriais para o Aumento da Competitividade da Indústria Brasileira: 5-Principais Segmentos Industriais; XII-Papel e Celulose. Brasília: Ministry of Development, Industry and Foreign Trade 2003. Available at: < http://www.mdic.gov.br/publica/sdp/acoes.html >. Accessed on: 21/03/03.

. Law 9.433, of 08 January 1997 - Institutes the National Water Resources Policy and Creates the National Water Resources Management System. Brasília: Federal Official Gazette, 09 January 1997.

. Law No. 6.938, of 31 August 1981 - Provides for national environmental policy, its purposes and mechanisms for formulation and application, and makes other provisions. Brasília: Federal Official Gazette of 2nd September 1981.

CÂMARA, Carla D. & LIMA, Walter de P. Clear-cutting a 50-year-old *Eucalyptus saligna* plantation: impacts on the water balance and water quality in an experimental watershed. Scientia Forestalis, IPEF, n. 56, p. 41-58, dec. 1999.

CÂMARA, Carla D. Effects of clear-cutting eucalyptus on water balance and nutrient cycling in an experimental watershed. Dissertation (Master's in Forestry Science) - University of São Paulo, Luiz de Queiroz College of Agriculture, Piracicaba, 1999.

CNA. Campo Futuro surveys production costs for six crops. Brasília, 22/07/2022. Available at: <https://www.cnabrasil.org.br/ noticias/campo-futuro-analisa-custos-de-producao-de-seis-culturas- 2>. Accessed on: 2024.

CARVALHO, Alexandre M. de. Valorisation of Eucalyptus grandis x Eucalyptus urophylla hybrid wood through the joint production of small dimension sawn timber, cellulose and firewood. Dissertation (Master's in Forestry Engineering) - University of São Paulo, Luiz de Queiroz College of Agriculture, Piracicaba, 2000.

CARVALHO, Humberto M. de. The Methodology of the Decision-Making Process. Linhares: Postgraduate course in Strategic Human Resources Management, FACCL, 1997a.

. Situation Analysis - Methodology of the Decision Process. Linhares: FACCL, Postgraduate course in Strategic Human Resources Management, handout, 1997.

CENIBRA. Contains institutional, financial, social and technical information. Available at: < http://www.cenibra.com.br >. Accessed on: 31/Jan/2003.

CHIARADIA, Tatiana. ALSTON Seminar - ISO 14001 - Environmental Management System. Itajubá: Alston Brasil, 2001.

Rio Declaration on Environment and Development. United Nations Conference on Environment and Development, UN, Rio de Janeiro 1992. Proceedings... [Brasília]: IBAMA, n.d. Available at: <

http://www.ibama.gov.br > . Accessed on: 12 August 2002.
DURLO, Miguel A. Forestry Roads. Santa Maria: University Federal de Santa Maria, Forestry Engineering Course, 1979. mimeographed.
FEPAM. Manual for Analysing Industrial Risks. Porto Alegre: FEPAM-Industrial Pollution Control Division, 2002. Available at: <http://www. fepam.rs.gov.br> . Accessed on: 18 January 2003.
FERREIRA, O. C. O futuro do carvão vegetal na siderurgia. Economia & Energia, n.21, July-August/2000.
FLORIANO, Eduardo P. Auditoria de Operações - Florestas Rio Doce S.A. São Mateus: Florestas Rio Doce S.A., 1998. internal document.
. Monitoring in the forestry company. São Mateus: 1998. Available at: <http://wwwsr.unijui.tche.br/ambienteinteiro>. Accessed on: 14/Feb/2003.
. Annual forestry operating budget - Pedro Canário Forestry Centre. Pedro Canário: Bahia Sul, 1990/91/92/93. internal document.
FURTADO, J. S. et al. Economic analysis of the implementation of *Eucalyptus urograndis*. Revista Agrotecnologia, Ipameri, v.13, n.1, p.37-49, 2022.
FSC-BRASIL. Contains institutional information and standards. Available at: <http://www.fsc.org.br>. Accessed on: 14/Feb-2003.
GALETI, Paulo E. Conservação do Solo - Reflorestamento - Clima, 2ª ed. Campinas: Instituto Campineiro de Ensino Agrícola, 1973.
GONÇALVES, José L. M. Recomendações de Adubação para Eucalyptus, Pinus e Espécies Típicas da Mata Atlântica. Piracicaba: Universidade de São Paulo, Escola Superior de Agricultura "Luiz de Queiroz", Department of Forestry Sciences, 1995. Forestry Documents No. 15.
GOUDARD Beatriz & SINAY, Maria C. F. de. Operational research techniques for evaluating project alternatives with polluting potential. *In*: XXXIII Brazilian Operational Research Symposium - SBPO, 6 to 9 November 2001, Campos do Jordão. Proceedings... Rio de Janeiro: Brazilian Society of Operational Research. Available at: < http://www.sobrapo.org.br/simposios/xxxiii/artigos/ 098-ST308.pdf>. Accessed on: 31/Jan/2003.
GROBER, Ulrich. Von Kursachsen Nach Rio - Ein Lebensbild über den Erfinder der Nachhaltigkeit Hannß Carl Edler von Carlowitz und die Wegbeschreibung eines Konzeptes - aus der Silberstadt Freiberg. Available at: <http://www.forschungsheim.de/fachstelle/arb_carl.htm>. Accessed on: 5/Nov/2002.
GRUPO INTERNATIONAL PAPER DO BRASIL. A market that never stops - National overview. Available at: http://www.internationalpaperdobrasil .com.br. Accessed on: 31/Jan/2003.
KLABIN PULP AND PAPER GROUP. Contains institutional, financial,

social and technical information. Available at: <http://www.klabin.com.br>. Accessed on 31/Jan/2003.
RIPASA PULP AND PAPER GROUP. Contains institutional, financial, social and technical information. Available at: <http://www.ripasa.com.br>. Accessed on 31/Jan/2003.
IBA. Ibá annual report 2023. Available at: <https://iba.org/datafiles/publicacoes/relatorios/relatorio-anual- iba2023-r.pdf>. Accessed on: 2023.
IBAMA. Contains institutional and technical information. Available at: <http://www.ibama.gov.br> . Accessed on: 04/Feb/2003.
IBGE. Production of Vegetable Extraction and Forestry, 2022. Available at: <https://www.ibge.gov.br/estatisticas/economicas/ agricultura-e-pecuaria/9105-producao-da-extracao-vegetal-e-da- silvicultura.html>. Accessed on: 2024.
VALE/SUZANO. II Meeting for the technical integration of CVRD/SUZANO forestry companies. Presentation of Bahia Sul...Teixeira de Freitas: Bahia Sul Celulose, 1993.
INFOMONEY. Suzano (SUZB3) and Klabin (KLBN11): What are the trends for pulp and the companies in 2024? 07/12/2023. Available at: <https://www.infomoney.com.br/ mercados/suzano-suzb3-e-klabin-klbn11-quais-as-tendencias-para- a-celulose-e-para-as-companhias-em-2024/>. Accessed 2024.
INTERNATIONAL ELECTROTECHNICAL COMMISSION (IEC). IEC/TR2 60479-1 - Effects of current on human beings and livestock - Part 1: General aspects. Geneva (Switzerland): IEC, 1994-09.
IP - INTERNATIONAL PAPER CORPORATION. Contains institutional, financial, social and technical information. Available at: <http://www.internationalpaper. com>. Accessed on: 28/Jan/2003.
IPEF. Contains institutional and technical information and articles. Available at: <http://www.ipef.br> . Accessed on: 13/Feb/2003.
. Watershed Monitoring and Modelling Programme (PROMAB). Available at at:
<http://www.ipef.br/promab/members/>. Accessed on: 18/04/2003.
JARCEL. Contains institutional, financial, social and technical information. Available at: <http://www.jari.com.br>. Accessed on 31/Jan/2003.
JERÔNIMO; Lúcia H.; FOELKEL, Celso E. B.; FRIZZO, Sonia M. B. Utilisation of anthraquinone in the production of Eucalyptus saligna sulphate pulp and its effects on bleaching. *In*: Congreso Ibero Americano de Investigacion en Celulosa y Papel, Iguazu (Argentina), 18 to 20 October 2000. Proceedings... Possadas: Universidad Nacional de Misiones, [2000].
OPTION JOURNAL. Wood prices skyrocket; there's even a shortage of inputs for industry. 18 September 2023. Available at:<https://www. jornalopcao.com.br/economia/preco-da-madeira-dispara-chega-a- falta-

insumos-para-industria-531444/>. Accessed on: 2024.
JUVENAL, Thais L. & MATTOS, René Luiz G. The forestry sector in Brazil and the importance of reforestation. Rio de Janeiro: BNDES, 2002.
KRAMER, Paul J. & KOSLOWSKI, T. Fisiologia das Árvores. Lisbon: Calouste Gulbenkian Foundation, 1972.
KRÜGNER, Tasso L. Eucalyptus Diseases. *In*: Manual de Fitopatologia II. São Paulo: CERES, 1980. pág. 275-296.
LANI, João L.; RESENDE, Mauro; REZENDE, Sérvulo B. de; KROB, Alexandre; FARIAS, Antônio P. Avaliação e Ações Prioritárias para a Conservação da Biodiversidade dos Biomas Mata Atlântica e Campos Sulinos. Viçosa: Centre for the Study of Planning and Land Use - NEPUT, 1999.
LIMA, Walter de P. Impacto ambiental do eucalipto. São Paulo: EDUSP, 1993.
MACEDO, A.R.P; VALENÇA, A.C.V. The paper industry in Brazil and worldwide: an overview. BNDES, 1994.
MACEDO, Angela R. P. & LEITE, Elizabete T. Forest Products - Pulp and market pulp - perspectives 1997 / 2001. Rio de Janeiro: BNDES, 2002.
MALINOVSKI, J. R. & CAMARGO, C. M. S. Eucalyptus cultivation in the Brazilian context. *In* : Revista da Madeira, Curitiba, Year 11, No. 59, September 2001.
MENDES, Moisés. Environmentalist José Lutzenberger once again provokes controversy, despite asthma-related health problems. Zero Hora, Porto Alegre, Sunday 14 April 2002.
MILLER, Kenton. Bioregional Planning. Brasília: IBAMA, Série Meio Ambiente Debate, n° 14, 1997. 28p.
MIRSHAWKA, Victor. The Implementation of Quality and Productivity by Dr Deming's Method. São Paulo: McGraw-Hill, 1990.
MOHR, R. R. Failure Modes and Effects Analysis. 8ª ed. St. Louis (USA): Jacobs Sverdrup, Feb. 2002.
MONTARDO, Doris K. Recovery of the Physical Environment - Environmental Management Policies. Santa Rosa: UNIJUÍ - Universidade Regional do Noroeste do Estado do Rio Grande do Sul, Postgraduate Course in Environmental Management, Transparencies presented in class, 2002.
MOREIRA, J. M. M. A. P. ; ZANCANARO, G. G.; VUADEN, E.; SIMIONI, F. J. Analysis of the financial profitability of a *Eucalyptus urophylla* S. T. Blake plantation for energy with funding from the ABC Programme. Colombo: Embrapa Florestas, 2022. Available at: <http://www.infoteca.cnptia.embrapa.br/handle/item/221>. Accessed on: 2024.
NEVES, M. R.; PORTO, M. The pulp and paper industry in Brazil. GUIDE - Pulp and Paper Manufacturers and Suppliers, 2023.
OLIVEIRA, José A. P. de & WAISSMAN, Vera. Integrating action and

communication for an environmental marketing strategy: the Aracruz Celulose case. Rio de Janeiro: Fundação Getúlio Vargas - FGV, Escola Brasileira de Administração Pública e de Empresas - EBAPE, 2002.

PARANÁ. Definition and Classification of Pesticides. State Health Department, Paraná Health Institute, Environmental Health Centre. Available at: < http://www.saude.pr.gov.br/ Saude_ambiental/Agrotoxicos/definicao.htm >. Accessed: 18/09/2002.

PEREIRA, Helio dos S. Instituciones Forestales - Brasil. Santiago (Chile): Proyecto: Informacion y analisis para el manejo forestal sostenible: integrando esfuerzos nacionales y internacionales en 13 paises tropicales en america latina (GCP/RLA/133/EC), mayo 2001.

PEREIRA, J.C.D.; STURION, J.A.; HIGA, A.R.; HIGA, R.C.V.; SHIMIZU, J.Y.. Wood characteristics of some eucalyptus species planted in Brazil. Colombo: Embrapa-Florestas, 2000. Documents no. 38.

PULP&PAPER International (PPI). Contains statistics, institutional, financial, social and technical information. Available at: <http://www.ppi-mfi.com>. Accessed on: 23/09/2002.

WOOD MAGAZINE. Curitiba: year 11, no. 64, May 2002.

. Eucalyptus - The wood of the future. Curitiba: Year 11, no. 59, September 2001.

RIZZINI. Carlos T. Useful trees and woods of Brazil - Manual of Brazilian dendrology. São Paulo: EDUSP-Edgar Blücher, 1971.

ROCHA, José S. M. da. Technical environmental education for primary, secondary and higher education. 2ª ed. Brasília: ABEAS, 2001.

SCARPINELLA, Gustavo D'Almeida. Reforestation in Brazil and the Kyoto Protocol. São Paulo, 2002. 182p. Dissertation (Interdisciplinary Master's Degree in Energy) - Inter-University Graduate Programme in Energy, University of São Paulo, São Paulo, 2002.

SILVEIRA, Ronaldo L. V. de A.; HIGASHI, Edson N.; SGARBI, Fábio; MUNIZ, Marta R. A. Be the doctor of your eucalyptus. Jornal Informações Agronômicas, Piracicaba, Potash and Phosphate Institute, n°93 - Arquivo do Agrònomo, N° 12, March 2001. insert.

SOARES, Sebastião R. EIA-RIMA/Licensing - Life cycle / Environmental gains / System evaluation - Part 10. Universidade Federal de Santa Catarina, Postgraduate programme in environmental engineering, 2002-b. Available at: < http://www.ens.ufsc.br/~soares > . Accessed on: 23/Nov/2002.

. Example of a matrix for assessing the significance of impacts. Florianópolis: Federal University of Santa Catarina, Department of Sanitary and Environmental Engineering, 2002-b. Available at: <http://www.ens. ufsc.br/~soares> . Accessed on: 23/Jan/2003.

. Environmental Indicators - Environmental Labelling. Florianópolis: Federal University of Santa Catarina, Department of Sanitary and Environmental Engineering, 2002. Available at: <http://www.ens. ufsc.br/~soares> . Accessed on: 23/Jan/2003.

. Environmental management system. Florianópolis: Federal University of Santa Catarina, Department of Sanitary Engineering and Ambiental, ENS 5125 - Gestão e Planejamento Ambiental, Aulas 06, 07 e 08, 2002-a. Available at: <http://www.ens. ufsc.br/~soares>. Accessed on: 23 January 2003.

SOARES, Thelma S. Optimising the use of wood in eucalyptus stands. Viçosa, 2002. 60 f. Dissertation (Master's Degree in Forest Science) - Postgraduate Programme in Forest Science, UFV, Viçosa, 2002-c.

BRAZILIAN FORESTRY SOCIETY - SBS. Contains institutional, statistical, social and technical information. Available at: <http://www.sbs. org.br> . Accessed on: 02/Feb/2003.

SOUSA, Roberto A. T. de M. Análise do fluxo de transporte rodoviário de toras curtas de eucalipto para algumas indústrias de celulose e de placas de composição no Estado de São Paulo. Piracicaba, 2000. 128 f. Thesis (Doctorate in Science) - ESALQ, USP, Applied Economics Area, Piracicaba, 2000.

SOUZA, Cintia R. de; SEIXAS, Fernando. Evaluation of different soil covers in erosion control on forest road slopes. Scientia Forestalis, Piracicaba, IPEF -, n. 60, p. 45-51, dec. 2001.

SOUZA, Helcio. The World Bank Group and global environmental management strategies for Brazil. Brasília: Inesc, 2001a. 188p.

SOUZA, Roberta L. de; OLIVEIRA, Meire J. L. de. Performance of the Market Pulp and Paper Industry: Brazil and Bahia - 1999 / 2001. Conjuntura e Planejamento, Salvador, SEI, n° 92, Jan. 2002. p. 19-29.

SUZANO. Contains institutional, financial, social and technical information. Available at: <http://www.suzano.com.br>. Accessed on 31/Jan/2003.

TAVARES JÚNIOR, João M. Uma aplicação da metodologia de análise do valor na verificação dos valores ambientais do processo produtivo numa empresa do setor cerâmico catarinense. Florianópolis: Federal University of Santa Catarina, PPGEP, 1997.

UOL-ECONOMIA/REUTERS. Suzano raises pulp prices between US$50 and US$80 from November. 17/10/2023. Available at: <https://economia.uol.com.br/noticias/reuters/2023/10/17/suzano- eleva-precos-de-celulose-entre-us50-e-us80-a-partir-de- novembro.htm>. Accessed on: 2024.

VCP - Votorantin Pulp and Paper Group. Contains institutional, financial, social and technical information. Available at: <http:// www.vcp.com.br>. Accessed on 31/Jan/2003.

VITAL, Ana Rosa T.; LIMA, Walter de P.; POGGIANI, Fabio; CAMARGO, Fausto R. A. de. Biogeochemistry of a micro-watershed after clear-cutting a 7-year-old eucalyptus plantation. Scientia Forestalis, Piracicaba, n° 55, p. 17-28, jun. 1999.

ZOBEL, B. J.; TALBERT, J. Aplied forest tree improvement. New York: John Wiley, 1984.

MIX
Papier aus verantwortungsvollen Quellen
Paper from responsible sources
FSC® C105338

Printed by Books on Demand GmbH, Norderstedt / Germany